Programmed Cell Death

Programmed Cell Death

Edited by

Yun-Bo Shi

National Institutes of Health
Bethesda, Maryland

Yufang Shi

American Red Cross
Rockville, Maryland

Yonghua Xu

Shanghai Institute of Cell Biology
Shanghai, China

and

David W. Scott

American Red Cross
Rockville, Maryland

Plenum Press • New York and London

Library of Congress Cataloging-in-Publication Data

International Symposium on Programmed Cell Death (1996 : Shanghai,
China)
    Programmed cell death / edited by Yun-Bo Shi ... [et al.].
        p.   cm.
    Includes bibliographical references and index.
    ISBN 0-306-45680-X
    1. Apoptosis--Congresses.   I. Shi, Yun-Bo.   II. Title.
QH671.I58   1996
571.9'39--dc21   97-22246
                                                            CIP

Proceedings of the International Symposium on Programmed Cell Death,
held September 8–12, 1996, in Shanghai, China

ISBN 0-306-45680-X

© 1997 Plenum Press, New York
A Division of Plenum Publishing Corporation
233 Spring Street, New York, N. Y. 10013

http://www.plenum.com

10 9 8 7 6 5 4 3 2 1

PREFACE

This volume contains papers that were presented and discussed at The 1996 International Symposium on Programmed Cell Death, which was held in the Shanghai Science Center of the Chinese Academy of Sciences on September 8–12, 1996.

Apoptosis has attracted great attention in the past several years. This is reflected in part by the exponential increase in the number of papers published on the subject. While several major scientific conferences have been held in recent years, this meeting represents the first major international scientific meeting on programmed cell death held in Asia, where fast economic growth promises a bright future for both basic and applied research in biomedical sciences. We organized the meeting with the belief that such a gathering would foster a closer interaction between scientists from the West and those in Asia.

Research on programmed cell death has expanded so extensively that no one meeting can cover all the important subjects related to apoptosis. The Shanghai meeting focused on several key areas ranging from well-established ones, such as cell death in the immune system, to emerging ones, such as the role of ECM in regulating cell fate. Specifically, the subjects presented and discussed included programmed cell death during development, the regulation and biochemical mechanisms of lymphocyte apoptosis, the involvement of extracellular matrix and its remodeling in programmed cell death, genes that cause or prevent cell death, and the application of apoptosis toward cancer therapy.

The meeting would not have been a success without the generous financial support from the National Institute of Child Health and Human Development, National Institutes of Health, USA; the Holland Laboratory of American Red Cross, USA; the National Natural Science Foundation of China; and Shanghai Institute of Cell Biology, Academia Sinica, China; as well as from our corporate sponsors. We are also deeply indebted to Dr. Arthur S. Levine, Scientific Director, National Institute of Child Health and Human Development, National Institutes of Health, USA, for his enthusiastic support and encouragement, and Dr. Rongxing Gan, the Secretary General of the meeting organization committee, whose effort made the symposium both scientifically productive and socially pleasant.

Y.-B. Shi, Bethesda, Maryland
Y. Shi, Rockville, Maryland
Y. Xu, Shanghai, China
D. W. Scott, Rockville, Maryland

SPONSORS

National Institute of Child Health and Human Development, USA
Holland Laboratory of American Red Cross, USA
National Natural Science Foundation of China
Shanghai Institute of Cell Biology, Academia Sinica, China
BASF Corp., USA
Genetech, Inc., USA
Shanghai Huaxin High Biotechnology, Inc., China
Amgen Center, USA
Beckman, USA
Boehringer Mannheim Corp., USA
DNAX Research Institute of Molecular and Cell Biology, USA
PharMingen, USA
Promega Corp., USA
Clontech Laboratories, USA
Hoffmann-LaRoche, Inc., USA
Life Technologies, Inc., USA
New England Biolabs., USA

CONTENTS

HOW HORMONES REGULATE PROGRAMMED CELL DEATH DURING AMPHIBIAN METAMORPHOSIS

Jamshed R. Tata

National Institute for Medical Research
The Ridgeway, Mill Hill
London NW7 1AA, United Kingdom

ABSTRACT

Extensive programmed cell death is initiated at the onset of amphibian metamorphosis, resulting in 100% of cells dying in some larval tissues, as during total regression of tail and gills. All cell death during metamorphosis is under the control of thyroid hormone (TH), which can initiate the process precociously in whole tadpoles, or in individual tissues in culture. The hormone prolactin (PRL), given exogenously, prevents natural and TH-induced metamorphosis. We have exploited this dual hormonal regulation in premetamorphic *Xenopus* tails in organ culture to identify and characterize "early" genes that are TH-induced and considered important for initiating cell death. Among the earliest genes activated by TH is that encoding the thyroid hormone receptor TRβ. This autoinduction of TR genes is considered important since, in blocking this process, PRL also inhibited the expression of other TH-inducible genes and prevented cell death. The expression of "early" genes other than TR genes, that are known to promote cell death (e.g. nur-77 and ICE) or survival (e.g. *Xenopus* bcl-2-like genes), is also considered to be important for the initiation of programmed cell death during amphibian metamorphosis. The possible significance of thyroid hormone-mediated amphibian metamorphosis to mammalian fetal development will be briefly discussed.

1. INTRODUCTION

Amphibian metamorphosis is an ideal model for studying programmed cell death (PCD) and apoptosis during vertebrate post-embryonic development (Tata, 1993, 1994). Among its several advantages are: a) many of the biochemical changes are similar to those seen during the acquisition of the adult phenotype at late embryonic or foetal stages of development in mammals; b) the process is obligatorily under thyroid hormonal control; c)

Programmed Cell Death, edited by Shi *et al.*
Plenum Press, New York, 1997

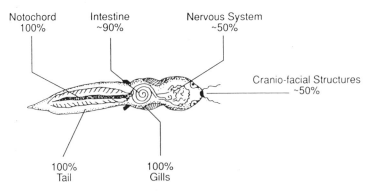

Figure 1. Schematic representation of the major tissues of an anuran tadpole undergoing regression during metamorphosis. The numbers indicate the extent of programmed cell death expressed as % of larval cells lost upon completion of metamorphosis.

the same hormonal signal activates different cell type-specific genetic programs in different tissues, and d) many aspects of metamorphic changes, including induction of cell death, can be reproduced in tissue culture.

Figure 1 schematically depicts the major tissues in the frog tadpole that undergo substantial or total regression during metamorphosis. The tail and gills undergo total regression (Beckingham Smith and Tata, 1976a; Yoshizato, 1989; Tata, 1994), while other tissues such as the small intestine and nervous system undergo major structural and functional reorganization which is accompanied by 50–90% loss of cells (Kollros, 1981; Atkinson, 1981; Ishizuya-Oka and Shimozawa, 1992). The regression of tadpole tail and intestine has been studied in organ culture, whereby PCD is induced by the addition of near-physiological amounts of thyroid hormone (Tata, 1966; Tata et al., 1991; Ishizuya-Oka and Shimozawa, 1991).

2. PROGRAM FOR METAMORPHIC CELL DEATH IS ESTABLISHED EARLY IN DEVELOPMENT

It has been well known for several decades that thyroid hormone can initiate metamorphosis precociously in every tissue of the amphibian larva, including those programmed for total regression (see Beckingham Smith and Tata, 1976a; Gilbert and Frieden, 1981; Gilbert et al., 1996). What was not known for some time was how early in development does the tadpole acquire competence to respond to the metamorphic stimulus of the hormone, until it was shown to be established much earlier than was assumed (Tata, 1968). Stage 45 *Xenopus laevis* tadpoles, which are well pre-metamorphic, exposed to exogenous thyroid hormone (3,3',5-triiodo-L-thyronine, T_3) showed morphological and biochemical signs of metamorphosis 4 days after their exposure to T_3. By stage 52 the eruption of hind limb buds was also manifested 4 days after hormone addition. These responses are elicited before the tadpole's thyroid gland is activated to secrete T_3 and T_4 (L-thyroxine), which occurs at stage 53–54 in *Xenopus* tadpoles (5–6 weeks post-fertilization) (Leloup and Buscaglia, 1977; Tata et al., 1993). Thus, the program for metamorphosis, including cell death, is laid down early in development, considerably before it would be normally put into operation.

3. ANALYSIS OF HORMONAL REGULATION OF PCD

In the mid-1960s three laboratories were able to prepare viable explants of *Xenopus* and *Rana* tadpole tails that were able to undergo regression upon the addition to T_4 or T_3 to the organ cultures (see Tata, 1971). By optimising variables such as culture conditions, dose of hormone, developmental stage of larvae, etc., it was possible to reproduce *in vitro* the rate of regression and sequential responses of different cell types seen during metamorphosis of the whole tadpole. T_3-induced regression of cultured tails was thus shown to be accompanied by the activation of lytic enzymes known to be involved with dying cells, such as cathepsin, collagenase and nucleases (Tata, 1966).

Until the time the above tadpole tail organ culture experiments were performed, it was generally believed that hormonal and other signals that triggered off PCD and tissue regression did so by activating latent lysosomal enzymes (see Weber, 1969). The tadpole tail culture system made it possible to test the involvement of *de novo* protein synthesis with the use of inhibitors of RNA and protein synthesis puromycin and cycloheximide (Weber, 1965; Tata, 1966). Figure 2 shows the kinetics of induction by T_3 of regression of cultured tails from premetamorphic *Xenopus* tadpoles. Measurement of DNA/tail established that the shortening of tails in culture was proportional to cell loss. The addition of actinomycin D (and also cycloheximide or puromycin) demonstrated that *de novo* RNA and protein synthesis is required for initiating and sustaining thyroid hormone-induced PCD. Later, other workers were able to reproduce the same phenomenon of prevention of hormonally regulated tissue regression in insect metamorphosis by blocking RNA and protein synthesis (Lockshin, 1981). Thus, and apparently quite paradoxically, cytotoxic agents which normally kill cells were found to protect the cells programmed to die against experimentally induced regression.

As regards the identity of proteins, or genes encoding them, that are critical for initiating PCD, it was not possible to achieve this goal in the days preceding the advent of monoclonal antibody and recombinant DNA technologies. Attempts to do so using laborious double or triple isotopic labelling techniques, combined with 2D-gel electrophoresis, ended in failure (Beckingham Smith and Tata, 1976b). Later, with the availability of specific recombinant DNA probes and highly sensitive procedures for quantifying newly synthesized RNA, in many cases of PCD during post-embryonic development, the signals activating the process may induce a few early gene products which would lead to a cas-

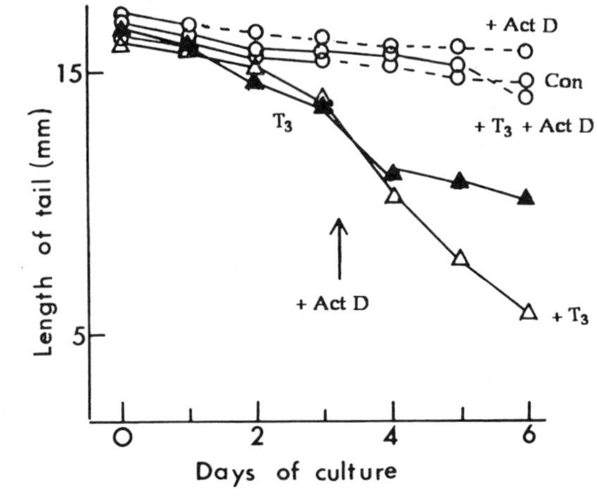

Figure 2. *De novo* RNA synthesis is required for PCD as shown by the inhibition by actinomycin D of T_3-induced regression of isolated *Xenopus* tadpole tails. 5×10^{-8}M T_3 and 5 μg/ml actinomycin D were added on day 0 of culture and the length of tails measured on days 1 to 6 of culture. See Tata (1966) for details.

cade of proteins to be synthesized sequentially leading to some which are components of the apoptotic process itself (see Ellis et al., 1995; Dexter et al., 1994).

Cytotoxic inhibitors of protein synthesis are of limited usefulness in extending our understanding of the mechanisms underlying PCD. Of greater value would be to exploit physiologically appropriate modulators of cell death. Our laboratory has hence exploited the anti-metamorphic properties of the hormone prolactin in further studying T_3-induced regression of *Xenopus* tails (Tata et al., 1991; Baker and Tata, 1992; Tata, 1994; Iwamuro and Tata, 1995). Exogenously administered prolactin has been known to block T_3-induced metamorphosis in many species of amphibia in whole tadpoles as well as in tail explants in culture (White and Nicoll, 1981; Ray and Dent, 1986).

Figure 3 illustrates an example of the inhibition by prolactin of T_3-induced regression of *Xenopus* tadpole tails in culture, which was monitored by measuring the reduction in tail length as well as DNA/cell (not shown). Figure 3 also shows that the synthetic glucocorticoid dexamethasone, which is known to potentiate the T_3 induction of metamorphosis (Galton, 1990; Kikuyama et al., 1993), boosted the effect of T_3. PRL also diminished the dexamethasone-potentiated induction of cultured tail regression by T_3. This effect highlights the importance of multiple hormonal interplay in postembryonic developmental processes.

How prolactin acts to inhibit T_3-induced metamorphosis is not known. We are just beginning to learn about the molecular basis of the action of this hormone characterized by a multiplicity of physiological action in almost all vertebrates (White and Nicoll, 1981; Rillema, 1987). Only recently has the prolactin receptor been characterized and it appears that the action of the receptor-ligand complex is transmitted intracellularly via the activation of JAK2 or *src* family of tyrosine kinases (Welte et al., 1994; Rui et al., 1994). As a consequence, the action of PRL would follow the JAK/STAT pathway of signal transduc-

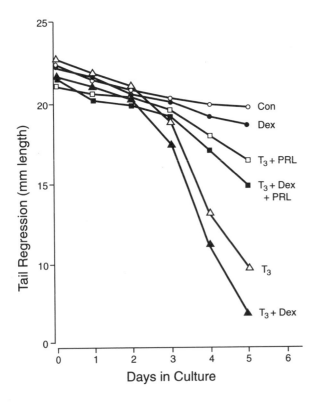

Figure 3. Effect of prolactin (PRL) and Dexamethasone (Dex) on the kinetics of *Xenopus* tadpole tail regression induced by T_3 in organ culture. Batches of 12 stage 53–55 *Xenopus* tadpole tails were set up in culture on day 0 and the different hormones were added on day 1. Regression was determined by daily measurement of tail length. Concentrations of hormones: T_3, 10^{-7}M; Dex, 10^{-7}M; PRL, 0.25 IU/ml. From Iwamuro and Tata, 1995.

tion leading to the phosphorylation-dephosphorylation of key regulatory proteins, including transcription factors. Some of the target proteins undergoing functionally relevant phosphorylation could be ubiquitous or specific transcription factors which could interfere with the ability of TR to upregulate TR itself. The interaction between PRL and T_3 can be considered as an example of "cross-talk" between hormonal signals. Much however remains to be done to establish if such a model could explain the antimetamorphic effect of PRL. Meanwhile, this hormone is a useful tool for experimentally exploring the various facets of PCD in a dual hormonally regulated system.

4. AUTOINDUCTION OF THYROID HORMONE RECEPTOR (TR) GENES

A central element in initiating the action of thyroid hormone is its receptor. For example, it is obvious that the molecular basis of this early metamorphic competence is most likely due to the well-advanced expression of thyroid hormone receptor (TR). As is now well known, the two TR genes in *Xenopus*, as in all vertebrates, are members of the steroid/thyroid hormone/retinoid receptor family, which are ligand activated transcription factors (Evans, 1988; Chatterjee and Tata, 1992). Most tissues in the pre-metamorphic *Xenopus* tadpole show the presence of TRα and β gene transcripts (Yaoita and Brown, 1990; Kawahara et al., 1991). It is of some significance in the context of cell death that TR mRNAs are detected in stage 44 *Xenopus* tadpole tail, as revealed by *in situ* hybridization (Kawahara et al., 1991). Of particular interest are the strong signals elicited in the tail tip, fin and notochord, which are just the regions of the tail that undergo substantial regression before muscle and skin tissue. Since the tadpole first shows a response to T_3 at stage 44/45, the detection of transcripts must also correspond to the presence of functional receptor. In fact, TR protein can be detected immunologically at early stages (Eliceiri and Brown, 1994).

Recent studies from the laboratories of Brown and Tata have shown that an autoinduction of TR mRNA, particularly TRβ mRNA, is an early response of all *Xenopus* tadpole tissues to T_3, irrespective of whether the tissues are programmed for morphogenesis, restructuring or complete regression (Yaoita and Brown, 1990; Kawahara et al., 1991; Baker and Tata, 1992; Tata, 1993, 1996; Tata et al., 1993). This upregulation by T_3 of its own receptor transcripts can also be mimicked in *Xenopus* cell lines (Kanamori and Brown, 1993; Machuca and Tata, 1992; Ulisse et al., 1996). It has recently been shown that the *Xenopus* TRβ gene promoter contains thyroid hormone responsive elements (TREs) whose activity can be regulated by liganded TR (Ranjan et al., 1994; Machuca et al., 1995). The phenomenon of autoinduction can be clearly seen in Figure 4 for both isoforms of TR genes in *Xenopus* tadpole tail culture. Under conditions in which PRL prevented T_3-induced cell loss (Figure 3), it also abolished the autoinduction of TRα and β genes. The same effect was seen in other tissues whether or not they undergo PCD during metamorphosis (Rabelo et al., 1994). Since TRβ is an "early" gene to respond to T_3 (Wang and Brown, 1993; Shi, 1996; Brown et al., 1996), it can be concluded, although not definitively proven, that autoinduction of thyroid hormone receptor is essential for initiation of PCD.

In common with some other members of the nuclear steroid/thyroid hormone/retinoic acid receptor family, thyroid hormone receptors (TRs) heterodimerize strongly with the 9-cis-retinoic acid receptors (RXRs) (see Mangelsdorf and Evans, 1995). Several different lines of evidence suggest that TRs exert their action as TR-RXR heterodimers binding to thyroid responsive element (TRE) in its DR+4 configuration, i.e. the hexanucleotide AGGTCA motif direct repeat separated by four nucleotides. Both *Xenopus* TRs (xTRα and xTRβ) heterodi-

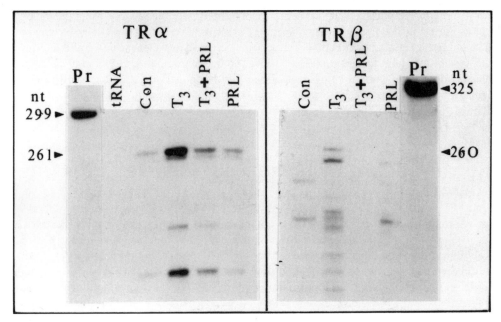

Figure 4. Inhibition by PRL of autoinduction of TRα and β mRNAs in organ cultures of stage 54 *Xenopus* tadpole tails. TRα and β transcripts were determined in total tail RNA by RNase protection assay. Other details are in Baker and Tata (1992).

merize with *Xenopus* RXRs (xRXRα, β and γ) to bind to DR+4 TREs in the promoters of TR target genes (Ranjan et al., 1994; Machuca et al., 1995; Wong and Shi, 1995; Ulisse et al., 1996). To understand better the role of RXR in the action of TR during amphibian metamorphosis, we recently studied the opposing actions of the glucocorticoid dexamethasone and prolactin in organ cultures of premetamorphic *Xenopus* tadpole tails induced to regress by T_3. It emerged that the three hormones elicited contrasting patterns of expression of xTRβ and xRXRγ (Iwamuro and Tata, 1995). In a dose- and time-dependent manner T_3 downregulated RXR mRNA accumulation, while dexamethasone had the opposite effect and PRL produced a more complex pattern of expression of the two nuclear receptors. In contrast to these findings, Wong and Shi (1995) have observed a coordinate enhancement by T_3 of expression of xTR and xRXR in whole *Xenopus* tadpole tissues. Hence, it became necessary to extend our earlier studies by transfection of *Xenopus* cell lines and tissues in organ culture with xTR and xRXR constructs to explain these findings in a functional context.

We have recently described the effect of T_3 and 9-cis-RA on the expression of xTRα and β and xRXRα and γ in *Xenopus* XTC-2 cells and the transcription from different TREs in XTC-2 transfected with expression vectors for xTRβ, xRXRα and xRXRγ. In another approach, experiments were carried out using a recently devised technique of microinjection of different expression and reporter DNA constructs into tadpole tail muscle *in vivo* before setting up organ cultures of tails to which T_3 was added (Ulisse et al., 1996). Our results show that xTR is the limiting factor in TH-induced transcription from TREs *in vivo* and *in vitro* but that overexpression of xRXRs did not significantly modify the transcriptional response. Furthermore, 9-cis-RA did not affect either the basal or T_3-induced transcription in XTC-2 cells, irrespective of overexpression of xRXRs or xTRβ. Our findings thus represent a novel facet of the functional implications of the interaction between two nuclear receptors that are heterodimeric partners (S. Ulisse, S. Iwamuro and J.R. Tata, unpublished).

5. EARLY GENES CONTROLLING CELL DEATH AND SURVIVAL

An ever-increasing number of genes have been identified in the last decade whose products are thought to play a major role in the process of programmed cell death (see Tomei and Cope, 1991; Ellis et al., 1991; Korsmeyer, 1992; Lavin and Watters, 1993; Dexter et al., 1994). These genes have been classified into two groups: a) those whose products induce cell death, such as *ced*-3 and -4 in nematodes and ICE (Interleukin 1B Converting Enzyme), *nur*-77, p53, c-*myc* (in the absence of serum), *bax* in vertebrates; b) those that confer cell survival, such as *ced*-9 in *C. elegans*, and *bcl*-2 and *bclx*$_L$ in vertebrates. Some of the genes that induce cell death have been considered to be "early" genes for the process, as for example, ICE, *nur*-77 and c-*myc*. Although some attempts have been made earlier to identify early genes that may be responsible for regression of intestine and tail during *Xenopus* metamorphosis (Shi and Brown, 1993; Wang and Brown, 1993), the expression of cell death or survival genes has been studied during amphibian metamorphosis only recently (Cruz-Reyes and Tata, 1995; Shi and Ishizuya-Oka, 1996; Shi, 1996; Brown et al., 1996; Shi, Y.-B., this volume).

In our attempts to identify and characterize both "early" cell death and cell survival genes in the context of PCD during amphibian metamorphosis, two *Xenopus* cDNAs were cloned, termed xR1 and xR11 (Cruz-Reyes and Tata, 1995). Their sequences were homologous to human cell survival gene *bcl*-x$_L$ (Boise et al., 1993). Constitutive expression of xR11 in rat fibroblasts conferred strong protection against the apoptotic effects of cytotoxic agents or the activation of c-*myc* in the absence of serum. However, both xR1 and xR11 continued to be expressed at as high a level in the regressing *Xenopus* tadpole tail during metamorphosis as in tissues undergoing morphogenesis or re-organization, such as limbs and brain. It is likely that during amphibian metamorphosis the expression of cell survival genes is not under developmental or hormonal control. We therefore turned our attention to early cell death genes, which include *nur*-77 and ICE, which are upregulated during T$_3$-induced regression of tadpole tails in culture, but it is not known if these are direct response genes (S. Iwamuro and J.R. Tata, unpublished). The reader is referred to recent publications from the laboratories of Shi and Brown which address in greater detail the question of early and/or direct response genes expressed during the regression of intestine and tail of metamorphosing *Xenopus* tadpoles (see Table 1 and Shi, 1996; Shi and Ishizuya-Oka, 1996; Brown et al., 1996).

As already mentioned, the autoinduction of *Xenopus* TRβ gene, which can be considered as an early and direct response gene in the regressing tail (Brown et al., 1996), is the most interesting phenomenon. In an attempt to clarify the role of TRβ gene upregulation in PCD, Ulisse et al. (1996) have recently taken the dominant-negative receptor approach. They show that, both in transfected *Xenopus* cell lines and pre-metamorphic

Table 1. Genes upregulated in *Xenopus* tadpole tail during metamorphosis

Transcription factors	Proteinases	ECM	Others
TRβ (DR)	stromelysin-3 (DR)	fibronectin	type III iodothyronine, deiodinase (DR)
Zn finger (BTEB) (DR)	collagenase 3	integrin α-1	unidentified -1 (DR)
bZip (E4BP4)	FAPα		unidentified-2 (DR)
bZip (Fra-2)	N-aspartyl dipeptidase		CRF-binding protein
			unidentified-3 (TS)

(DR), Direct response gene; (TS), tail-specific. Adapted from Brown *et al.* (1996).

Xenopus tadpole tails injected with various expression vectors, a dominant-negative *Xenopus* TRβ, as well as human homologs, totally abolished the upregulation of wild-type TRβ by T$_3$ (see Figure 5). We are now investigating, by a combination of biochemical and *in situ* hybridization techniques, the possibility that inhibition of TRβ autoinduction will

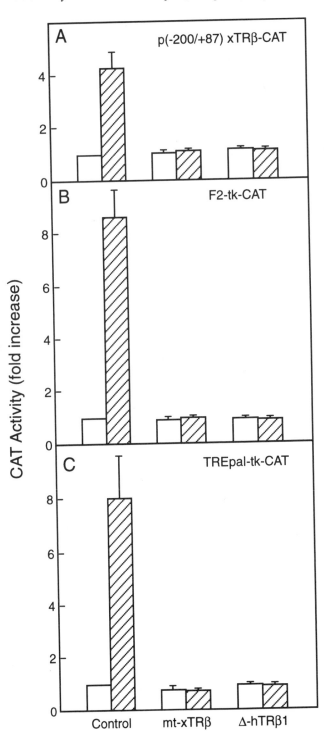

Figure 5. Full dominant-negative activities of the mutant human (mt-hTRβ) and Xenopus (xTRβ) TRβs on different TRE-CAT constructs *in vivo*. Tadpole tails were injected with 5 μg of p(−200/+87)xTRβ-CAT (A), F2-tk-CAT (B), or TREpal-tk-CAT (C) together with wt-xTRβ (1 μg) and pGL2 control (5 μg) in the absence or presence of either mt-xTRβ (10 μg) or -hTRβ1 (10 μg). The following day, the tadpole tails were excised and cultured for a further 5 days in the absence or presence of 10 nM T$_3$. At the end of the hormonal treatment, CAT activity in the tail extracts was measured as described. (Solid white), Basal; (striped), +T$_3$. Data are representative of two independent experiments. For further details, see Ulisse et al. (1996).

prevent the regression of the tadpole tail. If this turns out to be the case, then it can be concluded that the upregulation of thyroid hormone receptor is essential for the onset of cell death during metamorphosis.

6. CONCLUDING REMARKS

It is clear from the above review that tissue regression during amphibian metamorphosis (as also invertebrate metamorphosis) offers many advantages for investigating apoptotic cell death in post-embryonic development. The bi-directional control by hormones of PCD, i.e. both activation and arrest, the accurate mimicking of the physiological process in tissue culture, the precocious induction of cell death by a single hormone are some of the particularly attractive features, as illustrated here by the *Xenopus* tail regression model. It is the investigation of this system that was among the first to highlight the fact that *de novo* transcription and protein synthesis were essential for tissue regression.

More recently, the availability of techniques of recombinant DNA has made it possible to approach the important question of identifying and characterizing possible "early" genes that may play a pivotal role in determining the onset or prevention of apoptotic cell death. Consequently, an ever-growing number of genes thought to specify cell death or survival, such as in neurogenesis and immune cell turnover, have already been described, as is obvious from the contents of this volume. In the case of thyroid hormone-induced tadpole tail regression, the autoinduction of thyroid hormone receptor genes was found to be an early and essential event leading to cell death. A search for other likely "early" cell death and survival genes, such as the *ced*-3, *ced*-4, *nur*-77, ICE, *ced*-9, *myc* and the *bcl*-2 family, which control apoptotic process has been initiated for this amphibian system. A most challenging problem to be resolved, not only for post-embryonic tissue regression, but for all the different PCD processes, is to define whether or not an "early" gene plays a causal role leading to cell death or survival. Another important question is to characterize and explain the function of accessory factors that must play an important role in specifying tissue, gene and signal specificities.

ACKNOWLEDGMENTS

The work from our laboratory described in this review is the cumulative result of highly skilled effort of many investigators. I am particularly indebted to Mrs Betty Baker, Dr Kate Beckingham and Dr Akira Kawahara. I would also like to thank Mrs Ena Heather for the preparation of the manuscript.

REFERENCES

Atkinson, B.G. 1981. Biological basis of tissue regression and synthesis. *In* Metamorphosis. *Edited by* L.I. Gilbert and E. Frieden. Plenum Press, New York. pp. 397–444.

Baker, B.S. and Tata, J.R. 1992. Prolactin prevents the autoinduction of thyroid hormone receptor mRNAs during amphibian metamorphosis. Dev. Biol. 149, 463–467.

Beckingham Smith, K. and Tata, J.R. 1976a. The hormonal control of amphibian metamorphosis. *In* "Developmental Biology of Plants and Animals". *Edited by* C. Graham and P.F. Wareing. Blackwell, Oxford, pp. 232–245.

Beckingham Smith, K. and Tata, J.R. 1976b. Cell death. Are new proteins synthesised during hormone-induced tadpole tail regression? Exp. Cell Res. 100, 129–146.

Boise, L.H., Gonzáles-Garcia, M., Postema, C.E., Ding, L., Lindsten, T., Turka, L.A., Mao, X., Nunez, G. and Thompson, C.B. 1993. *bcl*-x, a *bcl*-2-related gene that functions as a dominant regulator of apoptotic cell death. Cell 7A, 597.

Brown, D.D., Wang, Z., Furlow, J.D., Kanamori, A., Schwartzman, R.A., Remo, B.F. and Pinder, A. 1996. The thyroid hormone-induced tail resorption program during *Xenopus laevis* metamorphosis. Proc. Natl. Acad. Sci. USA 93, 1924–1929.

Chatterjee, V.K.K. and Tata, J.R. 1992. Thyroid hormone receptors and their role in development. Cancer Surveys 14, 147–167.

Cruz-Reyes, J. and Tata, J.R. 1995. Cloning, characterization and expression of two *Xenopus bcl*-2-like cell-survival genes. Gene 158, 171–179.

Dexter, T.M., Raff, M.C. and Wyllie, A.H. (eds.) 1994. Death from inside out: the role of apoptosis in development, tissue homeostasis and malignancy. Phil. Trans Roy. Soc. B, Vol. 345.

Eliceiri, B. and Brown, D.D. 1994. Quantitation of endogenous thyroid hormone receptor α and β during embryogenesis and metamorphosis in *Xenopus laevis*. J. Biol. Chem. 269, 24459–24465.

Ellis, R.E., Yuan, J. and Horvitz, H.R. 1991. Mechanisms and functions of cell death. Ann. Rev. Cell Biol. 7, 663–698.

Evans, R.M. 1988. The steroid and thyroid hormone receptor superfamily. Science 240, 889–895.

Galton, V.A. 1990. Mechanisms underlying the acceleration of thyroid hormone-induced tadpole metamorphosis by corticosterone. Endocrinology 127, 2997–3002.

Gilbert, L.I. and Frieden, E. (eds.) 1981. "Metamorphosis: A problem in developmental biology". Plenum Press, New York.

Gilbert, L.I., Tata, J.R. and Atkinson, B.G. (eds.) 1996. "Metamorphosis". Academic Press, San Diego.

Ishizuya-Oka, A. and Shimozawa, A. 1991. Induction of metamorphosis by thyroid hormone in anuran small intestine cultured organotypically in vitro. In Vitro Cell. Dev. Biol. 27A, 853–857.

Ishizuya-Oka, A. and Shimozawa, A. 1992. Programmed cell death and heterolysis of larval epithelial cells by macrophage-like cells in the anuran small intestine in vivo and in vitro. J. Morphol. 213, 185–195.

Iwamuro, S. and Tata, J.R. 1995. Contrasting patterns of expression of thyroid hormone and retinoid X receptor genes during hormonal manipulation of *Xenopus* tadpole tail regression in culture. Mol. Cell. Endocrinol. 113, 235–243.

Kanamori, A. and Brown, D.D. 1993. Cultured cells as a model for amphibian metamorphosis. Proc. Natl. Acad. Sci. USA 90, 6013–6017.

Kawahara, A., Baker, B. and Tata, J.R. 1991. Developmental and regional expression of thyroid hormone receptor genes during *Xenopus* metamorphosis. Development 112, 933–943.

Kikuyama, S., Kawamura, K., Tanaka, S. and Yamamoto, K. 1993. Aspects of amphibian metamorphosis. Hormonal control. Int. Rev. Cytol. 145, 105–148.

Kollros, J.J. 1981. Transitions in the nervous system during amphibian metamorphosis. *In* Metamorphosis. *Edited by* L.I. Gilbert and E. Frieden. Plenum Press, New York. pp. 445–459.

Korsmeyer, S.J. 1992. Bcl-2 initiates a new category of oncogenes, Regulators of cell death. Blood 80, 879–886.

Lavin, M. and Watters, D. (eds.) 1993. The cellular and molecular biology of apoptosis. Harwood Academic Publishers, Chur.

Leloup, J. and Buscaglia, M. 1977. La triiodothyronine, hormone de la metamorphose des amphibiens. C.R. Acad. Sci. 284, 2261–2263.

Lockshin, R.A. 1981. Cell death in metamorphosis. In Cell death in biology and pathology. *Edited by* I.D. Bowen and R.A. Lockshin. Chapman and Hall, London, pp. 79–121.

Machuca, I. and Tata, J.R. 1992. Autoinduction of thyroid hormone receptor during metamorphosis is reproduced in *Xenopus* XTC-2 cells. Molec. Cell. Endocr. 87, 105–113.

Machuca, I., Esslemont, G., Fairclough, L. and Tata, J.R. 1995. Analysis of structure and expression of the *Xenopus* thyroid hormone receptor β (xTRβ) gene to explain its autoinduction. Molec. Endocrinol. 9, 96–107.

Mangelsdorf, D.J. and Evans, R.M. 1995. The RXR heterodimers and orphan receptors. Cell 83, 841–850.

Rabelo, E.M.L., Baker, B. and Tata, J.R. 1994. Interplay between thyroid hormone and estrogen in modulating expression of their receptor and vitellogenin genes during *Xenopus* metamorphosis. Mech. Develop. 45, 49–57.

Ranjan, M., Wong, J. and Shi, Y.-B. 1994. Transcriptional repression of *Xenopus* TRβ gene is mediated by a thyroid hormone response element located near the start site. J. Biol. Chem. 269, 24699–24705.

Ray, L.B. and Dent, J.N. 1986. Observations on the interaction of prolactin and thyroxine in the tail of the bullfrog tadpole. Gen. Comp. Endocrinol. 64, 36–43.

Rillema, J.A. (ed.) 1987. Actions of prolactin on molecular processes". Boca Raton, FL, CRC Press.

Rui, H., Lebrun, J.-J., Kirken, R.A., Kelly, P.A. and Farrar, W.L. 1994. JAK2 Activation and cell proliferation induced by antibody-mediated prolactin receptor dimerization. Endocrinology 135, 1299–1306.

Shi, Y.-B. 1996. Thyroid hormone-regulated early and late genes during amphibian metamorphosis. In: "Metamorphosis" (Gilbert, L.I., Tata, J.R. and Atkinson, B.G., eds.) pp. 505–538.

Shi, Y.-B. and Brown, D.D. 1993. The earliest changes in gene expression in tadpole intestine induced by thyroid hormone. J. Biol. Chem. 268, 20,312–20,317.

Shi, Y.-B. and Ishizuya-Oka, A. 1996. Biphasic intestinal development in amphibians. Embryogenesis and remodeling during metamorphosis. Curr. Topics in Dev. Biol. 32, 205–235.

Tata, J.R. 1966. Requirement for RNA protein synthesis for induced regression of the tadpole tail in organ culture. Dev. Biol. 13, 77–94.

Tata, J.R. 1968. Early metamorphic competence of *Xenopus* larvae. Dev. Biol. 18, 415–440.

Tata, J.R. 1971. Protein synthesis during amphibian metamorphosis. Current Topics Dev. Biol. 6, 79–110.

Tata, J.R. 1993. Gene expression during metamorphosis: An ideal model for post-embryonic development. BioEssays 15, 239–248.

Tata, J.R. 1994. Hormonal regulation of programmed cell death during amphibian metamorphosis. Biochem. Cell Biol. 72, 581–588.

Tata, J.R. 1996. Hormonal interplay and thyroid hormone receptor expression during amphibian metamorphosis. In "Metamorphosis" (Gilbert, L.I., Tata, J.R. and Atkinson, B.G. eds.) pp. 466–503. Academic Press, San Diego.

Tata, J.R., Kawahara, A. and Baker, B.S. 1991. Prolactin inhibits both thyroid hormone-induced morphogenesis and cell death in cultured amphibian larval tissues. Dev. Biol. 146, 72–80.

Tata, J.R., Baker, B.S., Machuca, I., Rabelo, E.M.L. and Yamauchi, K. 1993. Autoinduction of nuclear receptor genes and its significance. J. Steroid Biochem. Molec. Biol. 46, 105–119.

Tomei, L.D. and Cope, F.O. (eds.) 1991. The molecular basis of cell death. Cold Spring Harbor Laboratory Press, Cold Spring Harbor.

Ulisse, S., Esslemont, G., Baker, B.S., Chatterjee, V.K.K. and Tata, J.R. 1996. Dominant-negative mutant thyroid hormone receptors prevent transcription from *Xenopus* thyroid hormone receptor β gene promoter in response to thyroid hormone in *Xenopus* tadpoles *in vivo*. Proc. Natl. Acad. Sci. USA 93, 1205–1209.

Wang, Z. and Brown, D.D. 1993. Thyroid hormone-induced gene expression program for amphibian tail resorption. J. Biol. Chem. 268, 16,270–16,278.

Weber, R. 1965. Inhibitory effect of actinomycin on tail atrophy in *Xenopus* larvae at metamorphosis. Experientia 21, 665–666.

Weber, R. 1969. Tissue involution and lysosomal enzymes during anuran metamorphosis. *In* Lysosomes in biology and pathology. *Edited by* J.T. Dingle and H.B. Fell. Vol. I. North-Holland, Amsterdam. pp. 437–461.

Welte, T., Garimorth, K., Philipp, S. and Doppler, W. 1994. Prolactin-dependent activation of a tyrosine phosphorylated DNA binding factor in mouse mammary epithelial cells. Molec. Endocrinol. 8, 1091–1102.

White, B.A. and Nicoll, C.S. 1981. Hormonal control of amphibian metamorphosis. *In* Metamorphosis. A problem in developmental biology. *Edited by* L.I. Gilbert and E. Frieden. Plenum Press, New York. pp. 363–396.

Wong, J. and Shi, Y.-B. 1995. Coordinated regulation of and transcriptional activation by *Xenopus* thyroid hormone and retinoid X receptors. J. Biol. Chem. 270, 18479–18483.

Yaoita, Y. and Brown, D.D. 1990. A correlation of thyroid hormone receptor gene expression with amphibian metamorphosis. Genes & Dev. 4, 1917–1924.

Yoshizato, K. 1989. Biochemistry and cell biology of amphibian metamorphosis with a special emphasis on the mechanism of removal of larval organs. Int. Rev. Cytol. 119, 97–149.

ACTIVATION OF MATRIX METALLOPROTEINASE GENES DURING THYROID HORMONE-INDUCED APOPTOTIC TISSUE REMODELING

Yun-Bo Shi[1]* and Atsuko Ishizuya-Oka[2]†

[1]Laboratory of Molecular Embryology
National Institute of Child Health and Human Development
Bldg. 18T, Rm. 101
Bethesda, Maryland 20892-5430
[2]Department of Anatomy
Dokkyo University
School of Medicine
Mibu, Tochigi 321-02, Japan

ABSTRACT

Programmed cell death or apoptosis is an essential process during amphibian meta-morphosis, not only in the resorption of tadpole specific tissues but also in the remodeling of existing tissues and *de novo* development of adult ones. The control of metamorphosis by thyroid hormone has allowed the identification and characterization of a number of genes which are involved in this process. Among them are genes encoding matrix metallo-proteinases (MMPs), which are capable of digesting various components of the extracellu-lar matrix (ECM). Of particular interest is the MMP stromelysin-3. The gene is activated prior to larval cell death and its high levels of expression correlate with basal lamina (the ECM that separates epithelium and the mesenchyme) modification. These and other obser-vations implicate that stromelysin-3 may participate in ECM remodeling, which in turn regulates cell fate during metamorphosis. In addition, our analyses of the expression of several other MMPs indicate multiple MMPs function, in a tissue-dependent manner, at various steps of organ transformation, some in the regulation of cell death, proliferation, and differentiation, while others in post-apoptotic ECM removal.

* Correspondence: phone 301-402-1004; fax 301-402-0078; e-mail shi@helix.nih.gov
† Correspondence: phone 282-87-2124; fax 282-86-1463; e-mail i-oka@dokkyomed.ac.jp

Programmed Cell Death, edited by Shi *et al.*
Plenum Press, New York, 1997

INTRODUCTION

Selective cell elimination is a critical step during vertebrate organogenesis and tissue remodeling (Wyllie et al., 1980; Schwartzman and Cidlowski, 1993). Likely due to the necessity to have exact temporal and spatial regulation, developmental cell elimination occurs almost always through a process with precise genetic control, i.e., programmed cell death. In most of the cases, programmed cell death takes place with distinctive sequential morphological changes in a process also known as apoptosis.

One of the oldest known, most dramatic developmental cell elimination processes takes place during amphibian metamorphosis. This tadpole-to-frog transition involves systematic transformation of different organs (Dodd and Dodd, 1976; Gilbert and Frieden, 1981). Different organs undergo vastly different changes at different developmental stages. Two of the extreme transformations include limb development and tail resorption. In the first case, cells predetermined to become a limb undergo rapid proliferation and subsequent differentiation and morphogenesis to create a new structure. Developmentally, hind limb morphogenesis is one of the earliest metamorphic changes to takes place. In the other extreme, the tail, a tadpole specific organ comprising some of the same cell types (i.e., the connective tissue, muscle, and epidermis) as the developing limb, completely degenerates toward the end of metamorphosis. Most other organs, however, are present in both tadpoles and frogs. The metamorphic transition brings about partial but drastic transformations to these organs in order to prepare them for their physiological roles in the frog.

While complex, the entire process of amphibian metamorphosis is under the control of the thyroid hormone (Dodd and Dodd, 1976; Gilbert and Frieden, 1981). Furthermore, the regulation by thyroid hormone is organ autonomous as individual organs such as the limb, tail, and intestine, can be induced to metamorphose with thyroid hormone even when cultured individually *in vitro* (Dodd and Dodd, 1976; Tata et al., 1991; Ishizuya-Oka and Shimozawa, 1991). Thus, the hormone must control the expression of genes within each organ that are critical for the organ-specific transformation.

Indeed, thyroid hormone can regulate transcription directly through its nuclear receptors, the thyroid hormone receptors, which are DNA binding transcription factors (Yen and Chin, 1994; Tsai and O'Malley, 1994; Mangelsdorf et al., 1995; Shi et al., 1996). Many of the genes regulated by thyroid hormone in different metamorphosing tissues have been isolated and characterized (Gilbert et al., 1996; Brown, 1996). Of particular interests among them are genes encoding matrix metalloproteinases (MMPs). The temporal and spatial regulation of the MMP genes has implicated the involvement of MMPs in larval cell death and adult tissue development.

APOPTOSIS DURING METAMORPHOSIS

The resorption of larval tissues and the morphogenesis of many frog organs require the elimination of larval cells. The hormonal and genetic control of amphibian metamorphosis suggests that such cell removal occurs through programmed cell death. In fact, early observations by Kerr et al. (1974) have clearly demonstrated that tail reposition is accomplished through apoptosis.

Our own work focusses on the remodeling of the tadpole intestine during *Xenopus laevis* metamorphosis. The tadpole intestine comprises predominantly a single layer of larval epithelial cells that has only a single fold, the typhlosole, in the anterior one third of the small intestine (Marshall and Dixon, 1978; Ishizuya-Oka and Shimozawa, 1987a). Beginning around stage 56, the connective tissue begins to increase in thickness first in the typhlosole and then in the rest of the intestine by stage 60, when secondary epithelial cells proliferate as cell islets (McAvoy and Dixon, 1977; Ishizuya-Oka and Shimozawa, 1987a). Concurrently, the larval epithelial cells degenerate and are eventually replaced by the newly differentiated adult (secondary) epithelial cells (McAvoy and Dixon, 1977; Dauca and Hourdry, 1985; Yoshizato, 1989; Gilbert and Frieden, 1981). The differentiation of secondary epithelial cells between stages 63 and 66, the end of metamorphosis, is also accompanied by the formation of multiple intestinal folds, which are surrounded by elaborate connective tissue and muscles.

The degeneration of larval intestinal epithelium was first shown to be through apoptosis by electron microscopy both during natural metamorphosis and in organ cultures *in vitro* (Ishizuya-Oka and Shimozawa, 1992). This observation is confirmed by directly examining DNA fragmentation *in situ* using the TUNEL method (Ishizuya-Oka and Ueda, 1996). Thus, little or no DNA fragmentation is detectible in the larval intestine during pre- and pro-metamorphosis upto stage 59 when epithelial degeneration has not yet begun (Fig. 1A). In contrast, around stage 60, labeled cells suddenly increase in number only in the larval epithelum (Fig. 1B) indicating that nuclear DNA in those cells is fragmented, in agreement with the morphological observations. Finally, by the end of metamorphosis, labeled cells are restricted to the crests of the newly formed intestinal folds (Fig. 1C), demonstrating that the removal of adult intestinal epithelial cells also occurs through apoptosis.

ECM REMODELING DURING INTESTINAL METAMORPHOSIS

The extracellular matrix (ECM) serves as a structural support for the cells it surrounds and is thus essential for the integrity and morphology of different organs. Equally importantly, ECM can modulate a number of cellular functions including cell migration, proliferation, differentiation, and cell death (Hay, 1991; Schmidt et al., 1993; Ruoslahti and Reed, 1994).

The intestinal epithelium is separated from the connective tissue by a special ECM, the basal lamina, which is composed of laminin, entactin, collagens, and proteglycans, etc (Hay, 1991; Timple and Brown, 1996). As the epithelium undergoes metamorphic transformations, the basal lamina remodels concurrently. In premetamorphic *Xenopus laevis* tadpoles, the intestinal basal lamina is a thin, flat, and single-layered structure (Fig. 2a). Around stage 60, when adult epithelial cells begin to proliferate and larval epithelial cells begin to undergo apoptosis (Ishizuya-Oka and Ueda, 1996), the basal lamina lining both types of epithelia develops into a much thicker, multi-layered structure through extensive folding (Fig. 2b). The basal lamina lining the larval epithelium remains thick until the larval epithelium finally disappears, i.e. along with the massive epithelial apoptosis (Ishizuya-Oka and Shimozawa, 1987b; Murata and Merker, 1991). Interestingly, the basal lamina appears to be much more permeable at the climax of metamorphosis (stage 60–63) in spite of the increased thickness. This permeability is reflected by the frequently observed migration of macrophages across the ECM into the degenerating epithelium, where they participate in the removal of degenerated epithelial cells (Fig. 2c). In addition, extensive contacts are present between the proliferating adult epithelial cells and the fibroblasts on the other side of the basal lamina (Ishizuya-Oka and Shimozawa, 1987b).

Larval epithelial cell removal is essentially complete around stages 62–63. After stage 63, with the progress of intestinal morphogenesis, i.e. intestinal fold formation, the adult epithelial cells differentiate. Concurrently, the basal lamina becomes thin and flat again, underlining the differentiating adult epithelium.

In addition to the above changes in the ECM, extensive ECM degradation and synthesis also occur. In particular, the ECM underlining the apoptotic epithelial cells is eventually removed and new basal lamina synthesis is required as the multi-layered adult epithelial cells (cell islets) differentiate into a monolayer epithelium.

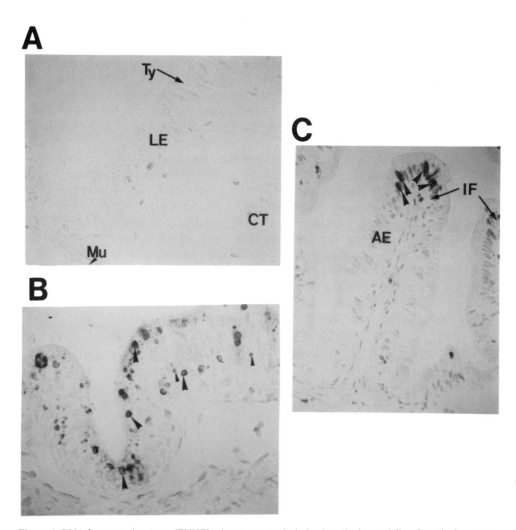

Figure 1. DNA fragmentation assay (TUNEL) detects apoptosis during intestinal remodeling. Intestinal cross sections from *Xenopus laevis* tadpoles at different stages were analyzed by the TUNEL assay (Gavrieli et al., 1992), by which cells undergoing apoptosis are labeled. (A) No cell death was observed in premetamorphic tadpole intestine at stage 59 (X630). (B) Most of the larval epithelial cells were labeled at the metamorphic climax (stage 60). Note the apoptotic morphology of the dying cells. Small arrow and large arrow heads refer to labeled nuclear fragments of apoptotic bodies and intact nuclei of the larval cells, respectively (X630). (C) Cell death (arrow heads) in post metamorphic frog intestine (stage 66) was limited to the crests of intestinal folds (IF), where the fully differentiated epithelial cells degenerate after functioning for a finite period of time (X420). Ty: typholosole; LE: larval epithelium; AE: adult epithelium; Mu; muscle; CT: connective tissue.

The intimate association of the basal lamina with surrounding cells, especially the epithelial cells, together with the close correlation of the ECM remodeling with intestinal transformation argues that ECM may play a critical role during metamorphosis. In particular, proper regulation of ECM remodeling is likely to be an important factor in epithelial cell fate determination, i.e. death vs proliferation and differentiation.

ACTIVATION OF MMP GENES DURING METAMORPHOSIS

Matrix metalloproteinases (MMPs) are extracellular enzymes that are capable of degrading various components of the ECM (Alexander and Werb, 1991; Woessner, 1991; Ma-

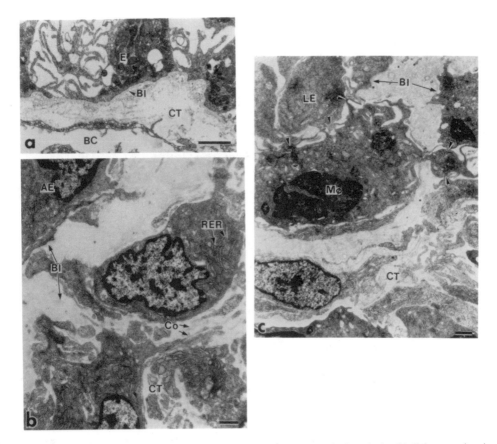

Figure 2. Electron microscopic examination reveals structural changes at the intestinal epithelial-connective tissue interface during metamorphosis. (a) A non-typhlosole region at stage 58. The basal laminal (Bl) is thin and lines the basal surface of the larval epithelium (E) (X12400). CT: connective tissue; BC: blood capillary. (b) At an early period (stage 61) of metamorphic climax. The basal lamina is thick due to vigorous folding. Fibroblasts just beneath the primordia of adult epithelium (AE) possess relatively developed rough endoplasmic reticulum (RER), and are surrounded by collagen fibrils (Co) (X7400). (c) Macrophage-like cells (Mϕ) migrate across the thick basal lamina into the degenerating larval epithelium at stage 62 (X6850). Bars: 1µm.

trisian, 1992; Birkedal-Hansen et al., 1993). This growing family of enzymes include collagenases, gelatinases, and stromelysins, etc., each of which has different but often overlapping substrate specificities (Sang and Douglas, 1996). They are secreted into the ECM as proenzymes with a possible exception of stromelysin-3 (ST3), which appears to be secreted in the active form (Pei and Weiss, 1995). The proenzymes can be activated in the ECM through the proteolytic removal of the propeptide. The mature enzyme has a catalytic domain at the N-terminal half of the protein, which contains a conserved Zn^{2+}-binding site. Once activated, these MMPs can degrade components of ECM. Thus differential expression and activation of various MMPs can result in specific remodeling of the ECM.

The participation of MMPs in amphibian metamorphosis was first implicated over 30 years ago by the drastic increase in collagen degradation activity in the resorbing tadpole tail (Gross et al., 1966). However, the cloning of frog MMP genes came only after two *Xenopus* MMP genes were identified as thyroid hormone-inducible genes from subtractive differential screens (Wang and Brown, 1993; Shi and Brown, 1993). The full length proteins encoded by these two genes share strong homology (≥ 64% identity) to mammalian collagenase-3 and stromelysin-3, respectively (Brown et al., 1996; Patterton et al., 1995) and are of similar sizes as their mammalian homologs, suggesting a functional conservation as well. Another frog MMP, the *Rana Catesbeiana* collagenase-1 was cloned by screening an expression cDNA library with an antiserum against purified *Rana* tail collagenase (Oofusa et al., 1994). Although the Rana collagenase-1 is much smaller than the mammalian collagenase-1 (384 vs 469 amino acids), it shares over 81% identity with the human collagenase-1, suggesting that it is likely to be the homolog of the human collagenase.

Correlation of *Xenopus* Stromelysin-3 Expression with Cell Death during Metamorphosis

We have studied extensively the spatial and temporal expression of the *Xenopus* stromelysin-3 (ST3) gene. We have chosen this gene because of the interesting expression profile of its human homolog, which is expressed in most, if not all, human carcinomas (Basset et al., 1990; Muller et al., 1993). Furthermore, both the human and mouse ST3 are expressed during development in tissues where cell death takes place (Basset et al., 1990; Lefbvre et al., 1992). These results suggest that ST3 is involved both in apoptosis and cell migration, the processes that also occur during frog intestinal remodeling.

As mention in the introduction, thyroid hormone (T_3) is the causative agent of frog metamorphosis. It controls tissue remodeling by affecting the transcription of genes involved in metamorphosis through thyroid hormone receptors. The *Xenopus* ST3 gene is one such thyroid hormone response gene. It can be precociously activated in all organs/regions of premetamorphic tadpoles if exogenous T_3 is added to the rearing water (Fig. 3A, Wang and Brown, 1993; Shi and Brown, 1993; Patterton et al., 1995). Furthermore, this regulation is direct and at the transcriptional level as it occurs even if new protein synthesis is blocked by protein synthesis inhibitors.

The developmental expression of ST3 mRNA correlates strongly with organ specific metamorphosis (Fig. 3B; Patterton et al., 1995). In the tail, the ST3 expression is low until around stage 62, when it is drastically up-regulated, coinciding with the onset of massive tail resorption through programmed cell death (Nieuwkoop and Fabor, 1956; Kerr et al., 1974). In contrast, the ST3 mRNA levels are much lower in the hind limb throughout limb development (Fig. 3B). However, even in this case, higher levels of the mRNA are present during limb morphogenesis when interdigital cells degenerate at stages 54–56, considerably earlier that tail resorption. Finally, high levels of ST3 mRNA are present during larval intestinal epithelial degeneration and adult cell proliferation (stages 60–62; Fig. 3B).

Thus, the ST3 expression is temporally correlated with the stages when cell death occurs in all these organs, and the levels of its mRNA appear to correlate with the extents of cell death in these organs (Fig. 3B; Patterton et al., 1995). More importantly, the activation of the ST3 gene occurs prior to cell death. Thus, in the intestine, high levels of ST3 mRNA is already present at stage 60 when larval apoptosis is first detected by TUNEL (Ishizuya-Oka and Ueda, 1996). Similarly, ST3 mRNA reaches high levels by stage 62 in the tail when massive tail resorption (length reduction) has yet to occur. These results suggest that ST3 is involved in regulating larval epithelial apoptosis. However, as adult epithelial proliferation in the intestine is rapid around stages 60–62 and differentiation also starts around stage 62, it is possible that ST3 may also be involved for adult epithelial development.

Interestingly, *in situ* hybridization analysis have revealed that ST3 is expressed in the fibroblastic cells adjacent to the epithelium, but not actually in the apoptotic cells of the intestine (Patterton et al., 1995). Thus, it is logical to assume that as a putative MMP, ST3 influences epithelial apoptosis by modifying ECM, in particular the basal lamina that

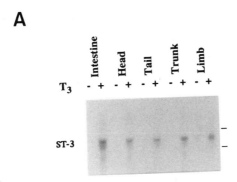

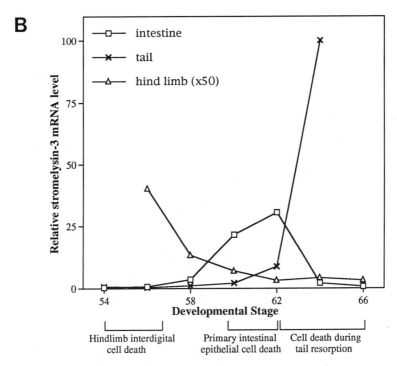

Figure 3. Stromelysin-3 (ST3) is regulated by thyroid hormone (T_3) and its expression correlates with cell death during metamorphosis. (A) ST3 can be activated by exogenous T_3 in different organs/regions of premetamorphic *Xenopus laevis* tadpoles (stage 52–54). Tadpoles were treated with 5nM T_3 and RNA from different organs was analyzed by Northern blot hybridization with a *Xenopus* ST3 cDNA probe. Each lane had 10 μg total RNA except the hindlimb lanes which had only 1.5 μg. The bars indicate the positions of 28S and 18S rRNA. (B) ST3 is expressed only during the period of cell death in the hindlimb, tail, and intestine. The relative mRNA levels were quantified by Northern blot hybridization (based on the data of Patterton et al., 1995). Note that ST3 mRNA levels in the hindlimb are about 50 time lower than those in the intestine and tail, probably corresponding to the much smaller fraction of the cells undergoing apoptosis in the limb.

separates the ST3-expressing fibroblasts and the dying epithelial cells. Consistent with this, detailed *in situ* hybridization analysis of the developmental ST3 expression has shown that ST3 expression is not only temporally but also spatially correlated with basal lamina modification (Ishizuya-Oka et al., 1996). In both pre- and post-metamorphic intestine, fibroblasts just beneath the thin and flat basal lamina do not express ST3. However, during metamorphosis, the intestinal basal lamina adjacent to ST3-expressing fibroblasts are thick, multiply folded, but more permeable as describe above. It is, therefore, possible that ST3 causes specific degradation/cleavage of certain ECM components that helps the folding of the ECM and results in the increased permeability. Such modifications may facilitate larval epithelial apoptosis and adult epithelial development. Conceivably, ST3 could play a similar role during limb morphogenesis and tail resorption. While such a function for ST3 remains to be proven, it is interesting to note that over-expression of the MMP stromelysin-1 in the mammary gland of transgenic mice leads to similar changes in basal lamina and alters mammary gland morphogenesis (Witty et al., 1995; Sympson et al., 1994).

Differential Regulation of Different MMPs during Metamorphosis

The multi-component nature of the ECM requires the participation of different MMPs in its remodeling and degradation, especially during tissue resorption. Thus, it was

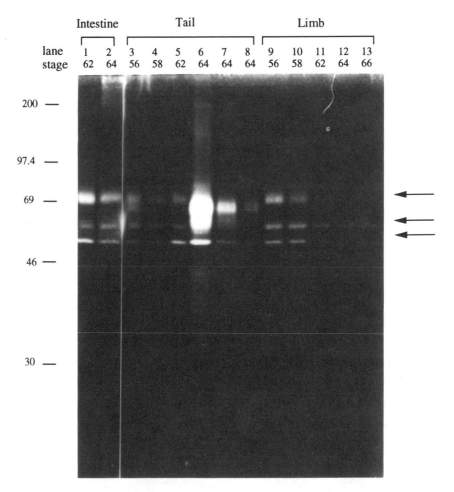

Figure 4A. Multiple MMPs are differentially involved in metamorphosis (cf. Figs. 4A and B). Gelatin zymography assay (McKerrow et al., 1985) for gelatin-degrading activity showing differential regulation of MMPs in different organs during metamorphosis (D. Patterton and Y.-B. Shi, unpublished). Each lane (except lanes 7 and 8) has 8μg total protein extracted from the intestine, tail, or hindlimb at the indicated developmental stage. Lanes 7 and 8 contained 1 and 0.25 μg stage 64 tail proteins, respectively. Protein size markers are indicated. Arrows point to the 74 kD, 61 kD and 56 kD gelatin degrading peptides.

not surprising to also identify the *Xenopus* collagenase-3 as a thyroid hormone-induced gene during tail resorption (Wang and Brown, 1993; Brown et al., 1996). To investigate the involvement of different MMPs during metamorphosis, we have used two different approaches. The first one makes use of the fact that many MMPs, especially gelatinases, can be renatured, after separating on a denaturing polyacrylamide gel, to become active toward a gelatin substrate preembeded in the gel (the gelatin zymography method, McKerrow et al., 1985). Thus, protein extracts from different organs at various metamorphic stages can be assayed by gelatin zymography to determine the relative levels of MMPs capable of degrading gelatin.

Using gelatin zymography, three species (74kD, 61kD and 56kD) were detected in the intestine, tail, and limb (Fig. 4A). In the intestine, like ST3, the levels of the 74kD and 56kD peptides are more abundant at stage 62 than at stage 64 and hence temporally correlated with the process of intestinal remodeling.

In the tail, both the 56 and 61kD enzymes increased markedly at stage 64, whereas the 74kD enzyme was replaced by an extremely high level of 67kD gelatin-degrading enzyme. Again like ST3, the levels of these MMPs were highly correlated with the process of tail resorption. In contrast, but similar to the lower levels of ST3 mRNA in the developing hindlimb during metamorphosis, much lower levels of all three putative MMPs were detected enzymatically at stage 56 and further reduced after stage 58. While it is difficult

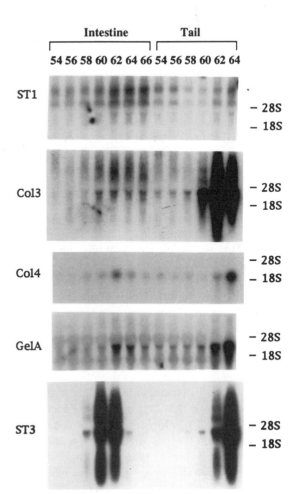

Figure 4B. Northern blot analyses demonstrate differential, organ-specific regulation of different MMP genes during *Xenopus* metamorphosis. Each lane had 10 µg total RNA except the tail lane at stage 64, which had only 5 µg. The probes used were: human stromelysin-1 (ST1, Wilhelm et al., 1987), *Xenopus* collagenase-3 (Col3, Wang and Brown, 1993), *Xenopus* collagenase-4 (Col4, Stolow et al., 1996), human gelatinase A (GelA, Collier et al., 1988), and *Xenopus* ST3 (Patterton et al., 1995).

to identify the MMPs producing these patterns of gelatinase activity, it is very unlikely that any of the three bands detected represent ST3. This is because they are all larger than the expected size of the frog ST3. Furthermore, the mouse ST3 has been shown to have only a weak gelatinase activity (Murphy et al., 1993). It is, however, tempting to speculate that the 67kD species represents the activated or truncated form of the 74kD enzyme and both are encoded by frog gelatinase A (GelA) based on the molecular weight of the mammalian GelA and the putative *Xenopus* GelA mRNA regulation patterns in the tail and intestine (Fig. 4B).

The second approach is to use homologous or heterologous cDNA probes to determine the mRNA levels for the corresponding MMPs. Overall, five genes have been analyzed due to the availability of appropriate cDNA probes. Three are homologous probes, encoding *Xenopus* ST3 (Patterton et al., 1995); collagenase-3 (Col3, Wang and Brown, 1993), and collagenase-4 (Col4, Stolow et al., 1996), respectively. The other two probes are the human cDNA clones for stromelysin-1 (ST1) and gelatinase A (GelA) (Patterton et al., 1995). The human GelA probe recognizes a *Xenopus* mRNA of similar size as the mammalian GelA mRNA while the ST1 probe hybridizes to two major RNAs of larger sizes than the mammalian ST1 mRNA (Patterton et al., 1995). However, both hybridization signals appear to be specific under the conditions used (Patterton et al., 1995), suggesting that the mRNAs detected likely encode the corresponding frog MMPs.

Unique but overlapping expression profiles have been observed for the five genes in the intestine and tail during metamorphosis (Fig. 4B, Patterton et al., 1995; Wang and Brown, 1993; Stolow et al., 1996). ST3 is the most up-regulated gene in both the intestine and tail during metamorphosis while ST1 has relatively constant mRNA levels throughout development in both organs. The two collagenase genes are up-regulated during tail resorption (stages 62–64, Fig. 4B). However, their mRNA levels are only a few fold higher in the metamorphosing intestine around stage 62 compared those at other stages (Fig. 4B). The putative *Xenopus* GelA has an expression profile that is most similar to that of ST3 (Fig. 4B). Its mRNA is at very low levels prior to metamorphosis but drastically up-regulated by stage 62 in both the intestine and tail. Subsequently, like ST3, the *Xenopus* GelA gene is repressed as intestinal remodeling is completed by the end of metamorphosis but remains high in the tail up to the end of tail resorption.

It is interesting to note that high levels of GelA mRNA in the intestine are reached later (at stage 62) that those of ST3 (at stage 60, Fig. 4B). This difference is also observed when premetamorphic tadpoles are treated with T_3 to induce precocious intestinal remodeling (Patterton et al., 1995). While the *Xenopus* ST3 gene is up-regulated very quickly (within a few hours) by the T_3 treatment, the GelA gene up-regulation is detectible only after a treatment of 3 days or longer. As described above, stage 62 is the time when intestinal epithelial cell death is mostly complete and adult epithelial differentiation is taking place. The activation of GelA and to a lesser extent, Col4, at this stage in the intestine suggests that these MMPs are involved in the removal of the ECM associated with the degenerated larval epithelium and remodeling of the ECM for adult epithelium differentiation. It also reenforces the idea that the earlier activation of the ST3 gene is important for the ECM remodeling that facilitates larval epithelial apoptosis. Such a role for ST3 may also exist in the tail, where ST3 gene also responds to exogenous T_3 treatment directly, within a few hours and much faster than the GelA gene (Patterton et al., 1995; Wang and Brown, 1993).

How ST3 influences tissue remodeling is yet unknown. Interestingly, mammalian ST3 is secreted in its enzymatically active form when over-expressed in tissue culture cells (Pei and Weiss, 1995). This has allowed the identification of one potential physiological substrate, the α1-proteinase inhibitor, a non-ECM derived serine proteinase inhibitor (Pei et al., 1994). Although, no natural ECM substrates for ST3 have been identi-fied, such substrates may still exist. On the other hand, the ability of ST3 to cleave a non-ECM substrate raises the possibility that ST3 may affect cell behavior through both ECM and non-ECM mediated pathways.

The other frog MMPs, i.e., ST1, GelA, Col3 and Col4, are expected to digest specific ECM substrates just like their mammalian counter parts, thus leading to the remodeling and degradation of the ECM during metamorphosis. In particular, the up-regulation of Col4 and GelA at stage 62 in the intestine when cell death is mostly completed implies that they may participate in ECM removal after cell death. It should be pointed out that direct demonstra-tion of the ECM degradation activity for these frog MMPs is lacking with the exception of Col4, which has been shown to be capable of cleaving native type I collagen similarly as human collagenase-1 (Stolow et al., 1996). However, the sequence conservation of the frog MMPs with the mammalian homologs argues for a conservation at the function level.

ECM AS A REGULATOR OF CELL FATE

The idea that ECM influences cell fate has been substantiated by recent experimen-tal evidence. One of the best studied system is the involution of the mammary gland. As the epithelial cells undergo post-lactation apoptosis, a number of MMP genes are activated (Lund et al., 1996; Talhouk et al., 1992). More importantly, by culturing the epithelial cells on different ECM matrices, it has been shown that the ECM can directly influence cell differentiation and survival (Boudreau et al., 1995; Roskelley et al., 1995). Similarly, ECM has been shown to be essential for the survival of several other types of cells and blocking the function of ECM receptor integrins can induce cell death (Bates et al., 1994; Brooks et al., 1994; Montgomery et al., 1994; Ruoslahti and Reed, 1994).

The mechanism for ECM to influence cellular function is still unknown. Clearly, one way to transduce the ECM signal into the cells is through cell surface ECM receptors, especially integrins (Ruoslahti and Reed, 1994; Brown and Yamada, 1995; Damsky and Werb, 1992). In the case of mammary gland development, it has been proposed that the interaction of ECM with its integrin receptors leads to the activation of focal adhesion ty-rosine kinase (FAK), which in turn transduces the signal through the MAP kinase pathway to the nucleus (Roskelley et al., 1995). This or similar mechanisms may be responsible for ECM-dependent gene transcription. It is presumably this ECM-regulated gene expression that is ultimately responsible for mediating the biological effects of MMPs in influencing cell fate in processes like intestinal metamorphosis and tail resorption.

ACKNOWLEDGMENT

We would like to thank Ms. T. Vo for preparing the manuscript.

REFERENCES

Alexander, C. M. and Werb, Z. (1991). Extracellular matrix degradation. In *Cell Biology of Extracellular Matrix*, 2nd ed. (ed. E. D. Hay), pp. 255–302. Plenum Press, New York.
Basset, P., Bellocq, J. P., Wolf, C., Stoll, I., Hutin, P., Limacher, J. M., Podhajcer, O. L., Chenard, M. P., Rio, M. C. and Chambon, P. (1990). A novel metalloproteinase gene specifically expressed in stromal cells of breast carcinomas. *Nature* 348, 699–704.

Bates, R.C., Buret, A., van Helden, D.F., Horton, M.A. and Burns, G.F. (1994) Apoptosis induced by inhibition of intercellular contact. *J. of Cell Biol.* **125**, 403–415.

Birkedal-Hansen, H., Moore, W. G. I., Bodden, M. K., Windsor, L. J., Birkedal-Hansen, B., DeCarlo, A., Engler, J. A. (1993) Matrix metalloproteinases: a review. *Crit. Rev. in Oral Biol. and Med.* 4(2) 197–250.

Boudreau, N., Sympson, C.J., Werb, Z. and Bissell, M.J. (1995) Suppression of ICE and apoptosis in mammary epithelial cells by extracellular matrix. *Science* 267 891–893.

Brooks, P.C., Montgomery, A.M.P., Rosenfeld, M., Reisfeld, R.A., Hu, T., Klier, G. and Cheresh, D.A. (1994) Integrin $a_v\beta_3$ antagonists promote tumor regression by inducing apoptosis of angiogenic blood vessels. *Cell* **79**, 1157–1164.

Brown DD, Wang Z, Furlow JD, Kanamori A, Schwartzman RA, Rmo BF, Pinder A. (1996) The thyroid hormone-induced tail resorption program during *Xenopus laevis* metamorphosis. *PNAS* **93**, 1924–1929

Brown, K.E. and Yamada, K.M. (1995) The role of integrins during vertebrate development. *Seminars in Develop. Biol.* **6**, 69–77.

Collier, I.E., Wilhelm, S.M., Eisen, A.Z., Marmer, B.L., Grant, G.A., Seltzer, J.L., Kronberger, A., He, C., Bauer, E.A. and Goldberg, G.I. (1988) H-ras oncogene-transformed human bronchial epithelial cells (TBE-1) secrete a single metalloprotease capable of degrading basement membrane collagen. *J. Biol. Chem.* **263**, 6579–6587.

Damsky, C.H. and Werb, Z. (1992) Signal transduction by integrin receptors for extracellular matrix: cooperative processing of extracellular information. *Current Biol.* **4**, 772–781.

Dauca, M. and Hourdry, J. 1985. Transformations in the intestinal epithelium during anuran metamorphosis. In Metamorphosis (Balls M. and Bownes, M., eds). pp. 36–58. The clarendon Press, Oxford, U.K.

Dodd, M. H. I. and Dodd, J. M. (1976). The biology of metamorphosis. In *Physiology of the Amphibia* (ed. B. Lofts), pp. 467–599. Academic Press, New York.

Gavrieli, Y., Sherman, Y. and Ben-Sasson, S.A. (1992) Ientification of programmed cell death in situ via specific labeling of nuclear DNA fragmentation. *J. Cell Biol.* 119, 493–501.

Gilbert, L.I. and E. Frieden. 1981. Metamorphosis: A problem in developmental biology, 2nd ed., Plenum Press, New York

Gilbert, L.I., Tata, J.R., Atkinson, B.G. (1996) Metamorphosis: Post-embryonic reprogramming of gene expression in amphibian and insect cells. Academic Press, New York.

Gross, J. (1966). How tadpoles lose their tails. *The journal of investigative dermatology.* 47, 274–277.

Hay, E.D. (1991). *Cell Biology of Extracellular Matrix*, 2nd ed. Plenum Press, New York.

Ishizuya-Oka, A. and Shimozawa, A. (1987a). Development of the connective tissue in the digestive tract of the larval and metamorphosing *Xenopus laevis*. *Anat.Anz. Jena.* 164, 81–93.

Ishizuya-Oka, A., Shimozawa, A. 1987b. Ultrastructural changes in the intestinal connective tissue of Xenopus laevis during metamorphosis. *J. Morphol.* **193**, 13–22

Ishizuya-Oka, A. and Shimozawa, A. (1991). Induction of metamorphosis by thyroid hormone in anuran small intestine cultured organotypically in vitro. *In vitro Cell Dev. Biol.* 27A, 853–857.

Ishizuya-Oka, A. and Shimozawa, A. (1992). Connective tissue is involved in adult epithelial development of the small intestine during anuran metamorphosis in vitro. *Roux's Arch Dev Biol.* **201**, 322–329.

Ishizuya-Oka, A., Shimozawa, A. 1992. Programmed cell death and heterolysis of larval epithelial cells by macrophage-like cells in the anuran small intestine in vivo and in vitro. *J. Morphol.* 213, 185–195.

Ishizuya-Oka, A. and Ueda, S. (1996) Apoptosis and cell proliferation in the Xenopus small intestine during metamorphosis. *Cell and Tissue Res.* **286**, 467–476.

Ishizuya-Oka, A., Ueda, S. and Shi, Y.-B. (1996) Transient expression of stromelysin-3 mRNA in the amphibian small intestine during metamorphosis. *Cell Tissue Res.* **283**, 325–329.

Kerr, J.F.R., Harmon, B. and Searle, J. (1974) An electron-microscope study of cell eletion in the anuran tadpole tail during spontaneous metamorphosis with special reference to apoptosis of striated muscle fibres. *J. Cell Sci.* **14**, 571–585.

Lefebvre, O., Wolf, C., Limacher, J. M., Hutin, P., Wendling, C., LeMeur, M., Basset, P. and Rion, M.C. (1992). The breast cancer-associated stromelysin-3 gene is expressed during mouse mammary gland apoptosis. *J. Cell Biol.* 119, 997–1002.

Lund, L.R., Romer, J., Thomasset, N., Solberg, H., Pyke, C., Bissell, M.J., Dono, K. and Werb, Z. (1996) Two distinct phases of apoptosis in mammary gland involution: proteinase-independent and -dependent pathways. *Development* **122**, 181–193.

Mangelsdorf DJ, Thummel C, Beato M, Herrlich P, Schutz G, Umesono K, Blumberg B, Kastner P, Mark M, Chambon P, Evans RM. (1995) The nuclear receptor superfamily: The second decade. *Cell* 83, 835–839.

Marshall, J. A. and Dixon, K. E. (1978). Cell specializaiton in the epithelium of the small intestine of feeding *Xenopus laevis*. *J. Anat.* 126, 133–144.

Matrisian, L. M. (1992). The matrix-degrading metalloproteinases. *Bioessays* 14, 455–463.

McAvoy, J. W. and Dixon, K. E. (1977). Cell proliferation and renewal in the small intestinal epithelium of metamorphosing and adult *Xenopus laevis*. *J. Exp. Zool.* 200, 129–238.

McKerrow, J. H., Pino-Heiss, S., Lindquist, R. and Werb, Z. (1985). Purification and characterization of an elastinolytic proteinase secreted by cercariae of Schistosoma mansoni. *J.Biol.Chem* 260, 3703–3707.

Montgomery, A.M.P., Reisfeld, R.A., Cheresh, D.A. (1994) Integrin $\alpha v \beta 3$ rescues melanoma cells from apoptosis in three dimensional dermal collagen. *Proc. Natl. Acad. Sci. USA* **91**, 8856–8860.

Muller, D., Wolf, C., Abecassis, J., Millon, R., Engelmann, A., Bronner, G., Rouyer, N., Rio, M. C., Eber, M., Methlin, G., Chambon, P. and Basset, P. (1993). Increased stromelysin 3 gene expression is associated with increased local invasiveness in head and neck squamous cell carcinomas. *Cancer Research* 53, 165–169.

Murata, E. and Merker, H.J. (1991) Morphologic changes of the basal lamina in the small intestine of Xenopus laevis during metamorphosis. *Acta Anat.* **140**, 60–69.

Murphy, G., Segain, J.-P., O'Shea, M., Cockett, M., Ioannou, C., Lefebvre, O., Chambon, P. and Basset, P. (1993) The 28-kDa N-terminal domain of mouse stromelysin-3- has the general properties of a weak metalloproteinase. *J. Biol. Chem.* **268**, 15435–15441.

Nieuwkoop, P. D. and Faber, J. (1956). *Normal table of Xenopus laevis*. North Holland Publishing, Amsterdam.

Oofusa, K., Yomori, S. and Yoshizato, K. (1994) Regionally and hormonally regulated expression of genes of collagen and collagenase in the anuran larval skin. Int. J. Dev. Biol. 38 345–350.

Patterton, D., Hayes, W.P. and Shi, Y.-B. (1995) Transcriptional activation of the matrix metalloproteinase gene stromelysin-3 coincides with thyroid hormone-induced cell death during frog metamorphosis. Dev. Biol. 167, 252–262

Pei, D. and Weiss, S.J. (1995) Furin-dependent intracellular activation of the human stromelysin-3 zymogen. *Nature* **375**, 244–247.

Pei, D., Majmudar, G. and Weiss, S.J. (1994) Hydrolytic inactivation of a breast carcinoma cell-derived serpin by human stromelysin-3. *The Journal of Biol. Chem.* **269**, 25849–25855.

Roskelley, C.D., Srebrow, A. and Bissell, M.J. (1995) A hierarchy of ECM-mediated signalling regulates tissue-specific gene expression. *Current Opinion in Cell Biol.* **7**, 736–747.

Ruoslahti, E. and Reed, J.C. (1994) Anchorage dependence, integrins, and apoptosis. *Cell* **77**, 477–478.

Sang, Q.A., Douglas, D.A. (1996) Computational sequence analysis of matrix metalloproteinases. J. Protein Chem. **15** 137–160.

Schmidt, J.W., Piepenhagen, P.A. and Nelson, W.J. (1993) Modulation of epithelial morphogenesis and cell fate by cell-to-cell signals and regulated cell adhesion. *Seminars in Cell Biol.* **4**, 161–173.

Schwartzman, R.A. and Cidlowski, J.A. (1993) Apoptosis: The biochemistry and molecular biology of programmed cell death. Endocrine reviews 14, 133–151.

Shi, Y.-B. and Brown, D. D. (1993). The earliest changes in gene expression in tadpole intestine induced by thyroid hormone. *J. Biol. Chem.* 268, 20312–20317.

Shi, Y.B., Wong, J. and Puzianowska-Kuznicka, M. (1996) Thyroid hormone receptors: Mechanisms of transcriptional regulation and roles during frog development. *J. Biomed. Sci.* **3**, 307–318.

Stolow, M.A., Bauzon, D.D., Li, J., Segwick, T., Liang, V.C-T., Sang, Q.A. and Shi, Y.-B. (1996) Ientification and characterization of a novel ollagenase in Xenopus laevis: Possible roles during frog development. *Mol. Biol. of the Cell* **7**, 1471–1483.

Sympson, C.J., Talhouk, R.S., Alexander, C.M., Chin, J.R., Clift, S.M., Bissell, M.J. an Werb, Z. (1994) Targeted expression of stromelysin-1 in mammary gland provides evidence for a role of proteinases in branching morphogenesis and the requirement for an intact basement membrane for tissue-specific gene expression. J. Cell Biol. 125, 681–693.

Talhouk, R.S., Bissell, M.J. and Werb, Z. (1992) Coordinated expression of extracellular matrix-degrading proteinases and their inhibitors regulates mammary epithelial function during involution. *Journal of Cell Biol.* **118**, 1271–1282.

Tata, J.R., A. Kawahara, and B.S. Baker. 1991. Prolactin inhibits both thyroid hormone-induced morphogenesis and cell death in cultured amphibian larval tissues. Devel. Biol. 146: 72–80

Timpl, R. and Brown, J.C. (1996) Supramolecular assembly of basement membranes. *BioEssays* **18**, 123–132.

Tsai M-J, O'Malley BW: 1994. Molecular mechanisms of action of steroid/thyroid receptor superfamily members. Ann Rev Biochem 63: 451–486

Wang, Z. and Brown, D. D. (1993). Thyroid hormone-induced gene expression program for amphibian tail resorption. *J. Biol. Chem.* 268, 16270–16278.

Wilhelm, S.M., Collier, I.E., Kronberger, A., Eisen, A.Z., Marmer, B.L., Grant, G.A., Bauer, E.A. and Goldberg, G.I. (1987) Human skin fibroblast stromelysin: Structure, glycosylation, substrate specificity, and differential expression in normal and tumorigenic cells. Proc. Natl. Acad. Sci. USA **84** 6725–6729.

Witty, J.P., Wright, J.H. and Matrisian, L.M. (1995) Matrix metalloproteinases are expressed during ductal an al-
    veolar mammary morphogenesis, and misregulation of stromelysin-1 in transgenic mice induces unsched-
    uled alveolar development. Mol. Biol. of the Cell 6, 1287–1303.

Woessner, J.F., Jr (1991). Matrix metalloproteinases and their inhibitors in connective tissue remodeling. FASEB
    J. 5 2145–2154.

Wyllie, A.H., Kerr, J.F.R., Curie, A.R. 1980. Cell death: the significance of apoptosis. Int. Rev. Cytol. 68: 251–306

Yen, P.M. and Chin, W.W. 1994. New advances in understanding the molecular mechanisms of thyroid hormone
    action. Trends Endocrinol. Metab. 5: 65–72

Yoshizato, K. 1989. Biochemistry and cell biology of amphibian metamorphosis with a special emphasis on the
    mechanism of removal of larval organs. Int. Rev. Cytol. 119: 97–149

3

MECHANISMS BY WHICH MATRIX METALLOPROTEINASES MAY INFLUENCE APOPTOSIS

William C. Powell and Lynn M. Matrisian

Department of Cell Biology
Vanderbilt University School of Medicine
Nashville, Tennessee 37232

1. INTRODUCTION

Apoptotic cell death plays a critical role in controlling cell number and tissue morphology in embryonic and adult organisms. Recently, there have been significant advances in the understanding of apoptosis at the cellular level. The cloning of diverse genes such as Fas and TNF α ligands/receptors, bcl-2, and ICE family members has lead to an increased knowledge of the signal transduction pathways used by cells to monitor and respond to their environment by inducing apoptosis. Components of the extracellular matrix play an important role in this process. Basement membrane and extracellular matrix contain information critical to the identity of the cell and the context in which it responds to external signals. Disruption of the matrix by the matrix-degrading metalloproteinases can therefore dramatically alter the response of cells to apoptotic signals. It is also interesting to consider that the proteolytic activity of metalloproteinases toward matrix components or other potential substrates provide alternative mechanisms by which degradative enzymes could actively influence apoptotic pathways. This review will discuss two systems in which metalloproteinases have been shown to be involved in apoptotic pathways and discuss potential mechanisms for this effect.

1.1. The Matrix Metalloproteinase Family

The matrix metalloproteinases (MMPs) are a class of secreted or transmembrane enzymes whose primary substrates are extracellular matrix proteins. Their enzymatic activity, which is optimal at physiological pH, requires a zinc ion in the active site and a second structural metal. MMPs can be inhibited by a family of proteins called tissue inhibitors of metalloproteinases (TIMPs), and recently a number of structure based synthetic inhibitors have been synthesized and are being tested for therapeutic applications (Brown, 1995).

Programmed Cell Death, edited by Shi *et al.*
Plenum Press, New York, 1997

The MMP family of enzymes currently contains 13 members that can be divided into 4 subcategories based on their protein domain structure (Powell and Matrisian, 1995 for review). The MMPs contain a "pre" signal sequence that directs the protein to the constitutive secretory pathway and they are secreted into the extracellular space in a latent or "pro" form with exceptions to be discussed below. The pro domain contains a highly conserved series of eight amino acids (PRCGVPDV) that function to maintain the enzyme in an inactive form. The cysteine in this sequence is unpaired and is bound to the zinc in the active site. Disruption of this interaction causes an intramolecular autocleavage that results in activated enzyme. The zinc in the active site is coordinated by the three histidines in the conserved sequence HEXGHXXGXXH. These three protein domains, pre, pro and catalytic, make up the core structure of the MMP family. The MMP matrilysin is the only member to contain only these domains.

The hemopexin domain MMPs, which include interstitial collagenase and stromelysin-1, have two additional domains: the hinge region and the hemopexin domain. The size of the hinge region is one of the determinants as to whether an MMP will degrade fibrillar collagens (Birkedal-Hansen et al., 1993 for review) The hemopexin domain, so named for its homology to the heme binding protein hemopexin, is a multifunctional domain that has been implicated in determining substrate specificity, inhibitor binding and cell surface activation. A subclass of transmembrane MMPs has been discovered recently and they are similar in structure to the hemopexin MMPs with the addition of a membrane spanning region at the COOH-terminus of the protein and the recognition site for furin between the pro and catalytic domains (Seiki, 1995 for review). This furin recognition site is of significant interest since furin is a member of the prohormone convertase family of proteins involved in intracellular protein processing (Hosaka et al., 1991), and has been shown to activate MT1-MMP (Sato et al., 1996). The MMP stromelysin-3 also contains the furin site and has been shown to be activated intracellularly in cells engineered to produce both stromelysin-3 and furin (Pei and Weiss, 1995). The fibronectin domain MMPs, which include gelatinase A and B, have an additional fibronectin-like domain which has homology to the collagen binding domain of fibronectin. This domain separates part of the catalytic site from the zinc binding sequence and has been shown to function in substrate binding. Gelatinase B has another domain that is homologous to the α 2 chain of type V collagen that could also function in substrate recognition.

To date research into the function of MMPs has focused on the ECM substrates of these enzymes. ECM substrates of MMPs can be grouped into three broad categories, fibrillar collagens, collagens and gelatins, and other components of the ECM and basement membrane. Fibrillar collagen is degraded by the collagenase MMPs while nonfibrillar collagen and gelatin is primarily degraded by the gelatinases and the stromelysins (Birkedal-Hansen et al., 1993 for review). Other matrix molecules such as proteoglycans, fibronectin, laminin, entactin, and tenascin are degraded by the stromelysin enzymes and matrilysin. The substrate specificities of the MMPs are not absolute and there is significant overlap between enzymes. Interestingly, some members of the MMP family cleave some non-ECM proteins, suggesting the possibility of alternative functions for certain members of the MMP family (Wilson and Matrisian, 1995).

1.2. MMPs Can Function in Protein Processing

One of the first non-ECM substrates described for an MMP was another MMP, when it was shown that digestion of activated interstitial collagenase with stromelysin-1 could "super" activate collagenase (Murphy et al., 1987). Since that time it has been demonstrated that many MMPs can activate one another (reviewed in Powell and Matrisian, 1995). Of particular interest is the identification of MT1-MMP, the first membrane bound MMP to be described. MT1-MMP was first cloned by homology based PCR (Sato et al., 1994) and was shown to be responsible for a previously described activity associated with cell membranes that was able to induce the activation of progelatinase A. Since then several more MT-MMPs have been cloned and all of them are able to activate progelatinase A (Puente et al., 1996; Takino et al., 1995, Will et al., 1995). Recent evidence indicates that MT1-MMP can degrade gelatins, but only after deletion of the transmembrane domain; it remains to be determined if MT1-MMP has ECM degrading ability while membrane bound (Imai et al., 1996). The lack of data on MT1-MMP matrix substrates, combined with its primary known function of activating gelatinase A and its intracellular activation by furin, leads one to consider the possibility that this enzyme plays a primary role in the proteolytic initiation of a cascade of degradative events.

The list of non-matrix molecule substrates of MMPs continues to grow. For example, matrilysin can cleave casein, insulin, transferrin, α 1-antitrypsin, urokinase type plasminogen activator (uPA), and tumor necrosis factor-α (TNF α) precursor (reviewed in Wilson and Matrisian, 1995). Matrilysin separates the catalytic portion of uPA from its receptor binding domain thus producing a soluble form that is not constrained to the cell surface. The receptor binding domain, which has homology to EGF, has been shown to induce cellular proliferation in some cell types (Arichini et al, 1994). The serine proteinase portion of uPA activates plasminogen to plasmin as well as functioning as a broad spectrum proteinase (Blasi et al., 1990). The potential affects of uPA cleavage can be contrasted with the ability of MMPs to release TNF α precursor from the cell surface (Gearing et al., 1994). The activity is similar, releasing a protein from the cell membrane, but the effects are quite different; TNF α can induce apoptosis in cells expressing the TNF α receptor (Baker and Reddy, 1996). Another member of the TNF α family, Fas ligand (FasL), has been shown to be released from the cell surface by an MMP-like activity (Kayagaki et al., 1995). FasL and its receptor, Fas, function in apoptosis in T-cells in particular and will be discussed in relationship to MMP-induced apoptosis below (Ju et al., 1995).

In this introduction we have stressed the dichotomy of MMP activities, degradation of ECM molecules and extracellular protein processing. These two seemingly different activities appear to converge in tissues undergoing active tissue remodeling. We will discuss the role MMPs play in the apoptotic process in two different animal models, transgenic expression of stromelysin-1 in mouse mammary gland and castration induced involution of the rodent prostate.

2. MMP FUNCTION IN THE MOUSE MAMMARY GLAND

Post-natal development of the mouse mammary gland begins at 5 weeks of age and continues through 13 weeks. During this time rudimentary ductal epithelium located at each nipple is stimulated by steroid hormones to begin branching morphogenesis. As development proceeds, stromelysin-1 and gelatinase A are produced by stromal cells surrounding the elongating glands (Witty et al., 1995a). To determine the potential role of stromelysin-1 during mammary gland branching morphogenesis, transgenic mice were made with stromelysin-1 under the transcriptional control of the MMTV-LTR which directs expression to mammary epithelial cells (Witty et al., 1995a). The stromelysin-1 cDNA used in the production of these mice was mutated in the conserved "pro" region. This change negates the need for activation of stromelysin-1 by other proteins (Sanchez-

Lopez et al., 1988). These transgenic mice, and a similar stromelysin-1 construct under the control of the whey acidic protein (WAP) promoter (Sympson et al., 1993), have a phenotype of increased number of lobular alveolar units when compared to wild type littermates. The mammary glands of stromelysin-1 transgenic mice appeared to have a morphology similar to a 10 day pregnant wild type mouse, displaying an increase in the number of epithelial cells and a differentiated phenotype (Sympson et al., 1993; Witty et al., 1995a). It has been established that mammary epithelial cell interaction with its basement membrane is important to its differentiation state and function (Lin and Bissell, 1993 for review). The basement membrane of the MMTV-stromelysin-1 transgenic mice appeared to be less organized than the basement membrane of nontransgenic littermates as determined by electron microscopy (Witty et al., 1995a), and immunohistochemistry revealed a loss of laminin and type IV collagen in lactating mammary glands expressing WAP-stromelysin-1 (Sympson et al, 1994). The authors conclude that the loss of intact basement membrane is likely to be responsible for the phenotypic changes observed.

In addition to displaying an effect of cellular proliferation and differentiation, stromelysin-1 transgenic mice also showed an increase in epithelial cell apoptosis. Witty and colleagues showed that there was an approximately 4-fold increase in apoptotic cells in MMTV-stromelysin-1 transgenic verses non-transgenic mammary glands at 12 to 13 weeks of age (Witty et al., 1995b), and Boudreau et al showed an increase in apoptotic cells in midpregnancy in WAP-stromelysin-1 mice (Boudreau et al, 1995). The degradation of the basement membrane by stromelysin-1 could also be responsible for the apoptotic effect on epithelial cells in these transgenic mice. Loss of contact with intact BM has been shown to send vascular endothelial cells into the apoptosis pathway (Meredith et al., 1993; Brooks et al, 1996). This hypothesis is supported by additional studies by Boudreau et al (1995; 1996) which show an effect of basement membrane in suppressing apoptosis in cultured mammary epithelial cells, and by Lund et al. (1996) in the lactating & involuting mammary gland. Thus, degradation of the basement membrane underlying the transgenic mammary epithelium by stromelysin-1 appears to result in the loss of essential survival factors and the induction of an apoptotic pathway.

An alternative to the view that it is a loss of intact basement membrane that induces apoptosis in the presence of metalloproteolytic activity is that degradation of the matrix may release factors or generate fragments that induce an apoptotic signal. For example, the matrix-associated factor Transforming Growth Factor-β has been suggested to induce apoptosis of liver cells (Schulte-Hermann et al, 1992). Fragments of matrix molecules may also act to induce programmed cell death. This has been demonstrated by the induction of apoptosis in melanoma cells by cyclic RGD peptides that have a high affinity for the $\alpha v \beta 3$ integrin (Mason et al., 1996) and cultured endothelial cells treated with RGD peptides also undergo apoptosis (Modlich et al., 1996). Thus, expression of a metalloproteinase may result in apoptosis as a result of either the loss of functional basement membrane, the release of positive-acting factors, or both.

3. MATRILYSIN IN PROSTATE

Matrilysin protein was originally purified from involuting post-partum rat uterus (Woessner and Taplin, 1988) and the association of MMP production in response to changing steroid hormone levels has been documented previously (for review see Hulboy et al 1997). The first study to address MMP expression in the prostate demonstrated that both matrilysin and gelatinase A were overexpressed in prostate cancer compared to normal prostate tissue from young healthy males (Pajouh et al., 1991). This study also showed that normal prostate epithelial cells express matrilysin in a highly focal expression pattern. The low levels of expression in the normal human prostate and the repression of matrilysin by steroid hormones suggested that matrilysin may play a functional role in post-castration involution of the rodent prostate.

Initially, it was hypothesized that metalloproteinases such as matrilysin could function to aid in extracellular matrix remodeling during involution of the prostate epithelium. Our first study demonstrated that matrilysin mRNA was induced following castration in the rat ventral prostate (RVP) with a time course of induction that correlated with the structural changes that occur late (days 4 through 8) during prostate involution (Powell et al., 1996). In contrast, western analysis for matrilysin protein showed that there was a low basal level of promatrilysin present in the prostate that was further induced following castration and peaked at day 3 post-castration. The matrilysin protein levels more closely correlated with the onset of apoptosis which occurs between 1 and 4 days following castration. Localization of matrilysin mRNA was performed by in situ hybridization and indicated that matrilysin was expressed focally in the epithelium of individual glands in the prostate. We have compared the localization of matrilysin to clusterin, a gene associated with tissues undergoing apoptotic involution (for review see Jenne and Tschopp, 1992). Clusterin production has recently been shown to protect cells from apoptosis rather than being involved in the apoptotic pathway (Sensibar et al., 1995). In contrast to the focal localization of matrilysin, clusterin mRNA was expressed in the majority of epithelial cells in all of the prostatic ducts and glands at 6 days post castration when most apoptotic cells have disappeared (unpublished data). This differential localization of matrilysin and clusterin lead us to hypothesize that matrilysin may have a functional role in apoptosis similar to that observed when stromelysin-1 was overexpressed in the mammary gland. Matrilysin produced by individual epithelial cells and the subsequent degradation of the underlying basement membrane could lead to apoptosis of selected cells.

Recent data from our laboratory indicates a potentially different mechanism by which matrilysin could contribute to apoptosis in the involuting prostate. We have produced 2 new antibodies directed against matrilysin; a rabbit polyclonal antibody that recognizes mouse matrilysin (Wilson et al., 1997) and a rat monoclonal antibody which detects human matrilysin (unpublished data). Immunohistochemistry using both of these antibodies shows that matrilysin is secreted to the apical surface of glandular epithelium including intestinal tumors (Wilson et al., 1997), seminal vesicle, breast and prostate (unpublished data). This was an unexpected finding considering that the primary substrates of matrilysin are believed to be extracellular matrix molecules that would be localized at the basal surface of the epithelium. If matrilysin is not degrading matrix molecules and aiding in tissue remodeling then what role might matrilysin play in the involuting prostate? As discussed earlier, metalloproteinases have been implicated in the release of both TNFα and FasL from cell surfaces based on data from synthetic inhibitors of MMPs and, in some cases, direct in vitro cleavage assays (Gearing et al., 1994; McGeehan et al., 1994, Kayagaki et al, 1995). Additional evidence that TNFα may play an important role comes from study done by Sensibar et al. (1995) showing that overexpression of clusterin in the prostate cancer cell line LnCaP protects these cells from TNFα induced apoptosis. This evidence leads to the hypothesis that TNFα or FasL play a role in post-castration apoptosis in the prostate and that matrilysin may facilitate this mechanism by proteolytic processing of the ligands. Our laboratory is beginning to address these questions in matrilysin deficient mice that we have recently described (Wilson et al., 1997). If this hypothesis is correct we would expect the matrilysin deficient mice to have reduced apoptosis in the prostate fol-

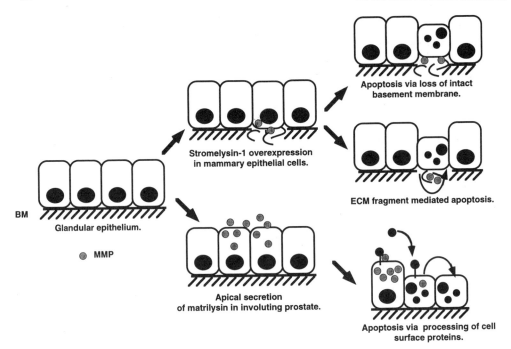

Figure 1. Potential mechanism by which MMPs induce apoptosis. Normal glandular epithelium has a close association with intact basement membrane. Altered expression of MMPs can disrupt this interaction such as in the
stromelysin-1 transgenic mice. The disruption of the interations between a cell and the basement membrane can
induce apoptosis via two different mechanisms. The loss of contact with intact basement membrane could in and
of itself intitiate apoptosis due to the cessation of survival signals from the matrix. Alternatively, the generation of
proteolytic fragments can unmask epitopes that can activly induce apoptosis in some cells. The apical secretion of
matrilysin presents an interesting and potential novel method by which MMPs can affect cellular processes. Matrilysin may process proteins on the lumenal surface of the cell that may have an effect on apoptosis in the rodent
prostate model.

lowing castration and potentially altered processing of TNFα or FasL. Thus, matrilysin
may have an unexpected function in prostate involution based on its trafficking pattern.

4. CONCLUSIONS

We have shown in two different model systems that MMPs are closely linked with
apoptosis. Stromelysin-1 and matrilysin appear to function in apoptosis via different
mechanisms. Stromelysin-1 overexpression in mammary epithelial cells may mediate its
effect through the degradation of basement membrane, a survival factor for many cell
types, so that in the absence of intact basement membrane the differentiated mammary
epithelial cells undergo apoptosis (Figure 1). The loss of basement membrane information
may produce a state of "amnesia", and the existing data suggest that cellular suicide is the
choice that many of these cells make. It is also possible that matrix degradation releases
matrix fragments, such as RGD-containing sequences, that play a role in inducing the signal transduction pathways of apoptosis. In contrast, matrilysin is induced in the prostate
following castration and is secreted on the apical surface of the epithelial cells, suggesting
that degradation of the basement membrane is not the primary function of matrilysin in
this model. Matrilysin may process or degrade lumenal surface proteins, potentially TNFα

or FasL, that function in the induction of apoptosis in prostate epithelial cells. These different mechanisms by which MMPs can affect apoptosis that may be related to a combination of differential secretion patterns and available substrates from either the basement membrane or the lumenal cell surface.

REFERENCES

Anichini, E., Fibbi, G., Pucci, M., Caldini, R., Chevanne, M., and Del Rosso, M. (1994). Production of second messengers following chemotactic and mitogenic urokinase-receptor interaction in human fibroblasts and mouse fibroblasts transfected with human urokinase receptor. Experimental Cell Research 231, 438–448.

Baker, S. J., and Reddy, E. P. (1996). Transducers of life and death: TNF receptor superfamily and associated proteins. Oncogene 12, 1–9.

Birkedal-Hansen, H., Moore, W. G. I., Bodden, M. K., Windsor, L. J., Birkedal-Hansen, B., DeCarlo, A., and Engler, J. A. (1993). Matrix Metalloproteinases: A Review. Critical Reviews in Oral Biology and Medicine 4, 197–250.

Blasi, F., Behrendt, N., Cubellis, M. V., Ellis, V., Lund, L. R., Masucci, M. T., Moller, L. B., Olson, D. P., Pedersen, N., Ploug, M., Ronne, E., and Dano, K. (1990). The urokinase receptor and regulation of cell surface plasminogen activation. Cell Differ.Dev. 32 , 247–254.

Boudreau, N., Sympson, C. J., Werb, Z., and Bissel, M. J. (1995). Suppression of ICE and apoptosis in mammary epithelial cells by extracellular matrix. Science 267, 891–893.

Boudreau, N., Werb, Z., and Bissell, M.J. (1996). Suppression of apoptosis by basement membrane requires three-dimensional tissue organization and withdrawal from the cell cycle. Proc. Natl. Acad. Sci. USA 93: 3509–3513.

Brooks, P.C., Montgomery, A.M.P., Rosenfeld, M., Reisfeld, R.A., Hu, T., Klier, G., and Cheresh D.A. (1996). Integrin αvβ3 antagonist promote tumor regression by inducing apoptosis of angiogenic blood vessels. Cell 79: 1157–1164.

Brown, P. D. (1995). Matrix metalloproteinase inhibitors: a novel class of anticancer agents. Advances in Enzyme Regulation 35, 293–301.

Gearing, A. J. H., Beckett, P., Christodoulou, M., Churchill, M., Clements, J., Davidson, A. H., Drummond, A. H., Galloway, W. A., Gilbert, R., Gordon, J. L., Leber, T. M., Mangan, M., Miller, K., Nayee, P., Owen, K., Patel, S., Thomas, W., Wells, G., Wood, L. M., and Woolley, K. (1994). Processing of tumour necrosis factor-α precursor by metalloproteinases. Nature 370 , 555–557.

Hosaka, M., Nagahama, M., Kim, W. S., Watanabe, T., Hatsuzawa, K., Ikemizu, J., Murakami, K., and Nakayama, K. (1991). Arg-X-Lys/Arg-Arg motif as a signal for precursor cleavage catalyzed by furin within the constitutive secretory pathway. Journal of Biological Chemistry 266, 12127–12130.

Hulboy, D.L., Rudolph, L.A., and Matrisian, L.M. (1996). Matrix metalloprotienases as mediators of reproductive function. Mol. Hum. Reprod. In Press

Imai, K., Ohuchi, E., Aoki, T., Nomura, H., Fugii, Y., Sato, H., Seiki, M., and Okada, Y. (1996). Membrane-type matrix metalloproteinase 1is a gelatinolytic enzyme and is secreted in a complex with tissue inhibitor of metalloproteinases 2. Cancer Res 56, 2707–2710.

Jenne, D. E., and Tschopp, J. (1992). Clusterin: the intriguing guises of a widely expressed glycoprotein. Trends Biochem Sci 17, 154–159.

Ju, S., Panka, D. J., Cui, H., Ettinger, R., El-Khatib, M., Sherr, D. H., Stanger, B. Z., and Marshak-Rothstein, A. (1995). Fas (CD95)/FasL interactions required for programmed cell death after T-cell activation. Nature 373, 444–448.

Kayagaki, N., Kawasaki, A., Ebata, T., Ohmoto, H., Ikeda, S., Inoue, S., Yoshino, K., Okumura, K., and Yagita, H. (1995). Metalloproteinase-mediated release of human Fas ligand. Journal of Experimental Medicine 182, 1777–1783.

Lin, C. Q., and Bissell, M. J. (1993). Multi-faceted regulation of cell differentiation by extracellular matrix. Faseb J 7, 737–743.

Lund, L. R., Romer, J., Thomasset, N., Solberg, H., Pyke, C., Bissell, M. J., Dano, K., and Werb, Z. (1996). Two distinct phases of apoptosis in mammary gland involution: proteinase-independent and -dependent pathways. Development 122, 181–93.

Marcotte, P. A., Kozan, I. M., Dorwin, S. A., and Ryan, J. M. (1992). The matrix metalloproteinase pump-1 catalyzes formation of low molecular weight (pro)urokinase in cultures of normal human kidney cells. Journal of Biological Chemistry 267, 13803–13806.

Mason, M. D., Allman, R., and Quibel, M. (1996). Adhesion molecules in melanoma:more than just superglue? J Royal Acad Sci 89, 393–395.

McGeehan, G. M., Becherer, J. D., Bast, R. C., Jr., Boyer, C. M., Champion, B., Connolly, K. M., Conway, J. G., Furdon, P., Karp, S., Kidao, S., McElroy, A. B., Nichols, J., Pryzwansky, K. M., Schoenen, F., Sekut, L., Truesdale, A., Verghese, M., Warner, J., Ways, J. P. (1994). Regulation of tumour necrosis factor-alpha processing by a metalloproteinase inhibitor. Nature 370, 558–561.

Meredith, J. E., Fazeli, B., and Schwartz, M. A. (1993). The extracellular matrix as a cell survival factor.. Mol.Biol.Cell 4 , 953–961.

Modlich, U., Kaup, F. J., and Augustin, H. G. (1996). Cyclic angiogenesis and blood vessel regression in the ovary: blood vessel regression during luteolysis involves endothelial cell detachment and vessel occlusion. Lab Invest 74, 771–780.

Murphy, G., Crockett, M. I., Stephens, P. E., Smith, B. J., and Docherty, A. J. P. (1987). Stromelysin is an activator of procollagenase - a study with natural and recombinant enzymes. Biochem. J. 248, 265–268.

Pajouh, M., Nagle, R., Breathnach, R., Finch, J., Brawer, M., and Bowden, G. (1991). Expression of metalloproteinase genes in human prostate cancer. J Cancer Res Clin Oncol 117, 144–150.

Pei, D., and Weiss, S. J. (1995). Furin-dependent intracellular activation of the human stromelysin-3 zymogen. Nature 375, 244–247.

Powell, W., and Matrisian, L. (1995). Complex roles of matrix metalloproteinases in tumor progression. In Attempts to understand metastasis formation: metastasis related molecules, U. Gunthert and W. Birchmeier, eds. (Berlin: Springer-Verlag), pp. 1–22.

Powell, W. C., Dormann, F. E., Jr., Michen, J. M., Matrisian, L. M., Nagle, R. B., and Bowden, G. T. (1996). Matrilysin expression in the involuting rat ventral prostate. The Prostate, 159–168.

Puente, X. S., Pendas, A. M., Llano, E., Velasco, G., and Lopez-Otin, C. (1996). Molecular cloning of a novel membrane-type matrix metalloproteinase from a human breast carcinoma. Cancer Research 56, 944–949.

Sanchez-Lopez, R., Alexander, C. M., Behrendtsen, O., Breathnach, R., and Werb, Z. (1993). Role of zinc-binding- and hemopexin domain-encoded sequences in the substrate specificity of collagenase and stromelysin-2 as revealed by chimeric proteins. J Biol Chem 268, 7238–7247.

Sanchez-Lopez, R., Nicholson, R., Gesnel, M. C., Matrisian, L. M., and Breathnach, R. (1988). Structure-function relationships in the collagenase family member transin. J Biol Chem 263, 11892–11899.

Sato, H., Takino, T., Okada, Y., Cao, J., Shinagawa, A., Yamamoto, E., and Seiki, M. (1994). A matrix metalloproteinase expressed on the surface of invasive tumour cells. Nature 370, 61–64.

Sato, H., Kinoshita, T., Takino, T., Nakayama, K., and Seiki, M. (1996). Activation of a recombinant membrane type 1-metalloproteinase (MT1-MMP) by furin and its interaction with tissue inhibitor of metalloproteinases (TIMP)-2. FEBS Letters *393*, 101–104.

Schulte-Hermann, R., Bursch, W., Kraupp-Grasl, B., Oberhammer, F., and Wagner, A. (1992). Programmed cell death and its protective role with particular reference to apoptosis. Toxicology Letters. *64–65 Spec No*, 569–574.

Seiki, M. (1995). Membrane type-matrix metalloproteinase and tumor invasion. In Attempts to understand metastasis formation: metastasis related molecules, U. Gunthert and W. Birchmeier, eds. (Berlin: Springer-Verlag), pp. 23–32.

Sensibar, J. A., Sutkowski, D. M., Raffo, A., Buttyan, R., Griswold, M. D., Sylvester, S. R., Kozlowski, J. M., and Lee, C. (1995). Prevention of cell death induced by tumor necrosis factor alpha in LNCaP cells by overexpression of sulfated glycoprotein-2 (clusterin). Cancer Res 55, 2431–2437.

Sympson, C. J., Alexander, C. M., Chin, J. R., Werb, Z., and Bissell, M. J. (1993). Transgenic expression of stromelysin from the WAP promoter alters branching morphogenesis during mammary gland development and results in precocious expression of milk genes. J. Cell Biol.125: 681–693.

Takino, T., Sato, H., Shinagawa, A., and Seiki, M. (1995). Identification of the second membrane-type matrix metalloproteinase (MT-MMP-2) gene from a human placenta cDNA library. J. Biol. Chem 270, 23013–23020.

Will, H., and Hinzmann, B. (1995). cDNA sequence and mRNA tissue distribution of a novel human matrix metalloproteinase with a potential transmembrane segment. Eur. J. Biochem. *231*, 602–608.

Wilson, C., and Matrisian, L. (1996). Matrilysin: an epithelial matrix metalloproteinase with potentially novel functions. International Journal of Biochemistry and Cell Biology 28: 123–136.

Wilson, C. L., Heppner, K. J., Labosky, P. A., Hogan, B. L. M., and Matrisian, L. M. (1997). Intestinal tumorigenesis is suppressed in mice lacking the metalloproteinase matrilysin. Proc. Natl. Acad. Sci. USA. In Press.

Witty, J. P., Wright, J. H., and Matrisian, L. M. (1995a). Matrix metalloproteinases are expressed during ductal and alveolar mammary morphogenesis, and misregulation of stromelysin-1 in transgenic mice induces unscheduled alveolar development. Molecular Biology of the Cell 6, 1287–303.

Witty, J. P., Lempka, T., Coffey, R. J., Jr., and Matrisian, L. M. (1995b). Decreased tumor formation in 7,12-dimethylbenzanthracene-treated stromelysin-1 transgenic mice is associated with alterations in mammary epithelial cell apoptosis. Cancer Research 55, 1401–1406.

Woessner, J., and Taplin, C. (1988). Purification and properties of a small latent matrix metalloproteinase of the rat uterus. J Biol Chem 263, 16918–16925.

4

THE ROLE OF INTEGRIN αVβ3 IN CELL SURVIVAL AND ANGIOGENESIS

Staffan Strömblad, Peter C. Brooks, Jürgen Becker, Mauricio Rosenfeld, and
David A. Cheresh

Departments of Immunology and Vascular Biology
The Scripps Research Institute
10666 N. Torrey Pines Rd.
La Jolla, California 92037

1. ANGIOGENESIS

The outgrowth of vasculature from pre-existing blood vessels is known as angiogenesis. The formation of new blood vessels facilitates physiological processes of embryonic development, the female reproductive cycle and wound-healing (Folkman (1995)). However, deregulated angiogenesis plays a critical role in various pathological mechanisms such as solid tumor formation, metastasis, childhood hemangiomas, diabetic retinopathy, macular degeneration, psoriasis and in inflammation related diseases such as rheumatoid arthritis, osteoarthritis and ulcerative colitis (Folkman (1995)). In particular, the expansion of solid tumors beyond a minimal size is critically dependent on neovascularization to supply oxygen, nutrients and growth factors. Angiogenesis is also crucial for the formation of metastases at secondary sites and accordingly, the degree of vascularization of certain tumors is correlated with a poor clinical prognosis and an increased risk of metastasis (Wedner (1995)).

The mechanism of tumor-induced angiogenesis can be generally divided into three phases. First, angiogenic stimulators, such as growth factors are released from tumors and/or inflammatory cells. A wide range of potential stimulators of angiogenesis have been identified. In particular, basic fibroblast growth factor (bFGF) and vascular endothelial growth factor (VEGF) released from various tumors have been suggested to be crusial for the induction of angiogenesis. These factors stimulate vascular cell proliferation and invasive behavior thereby promoting blood vessel growth and invasion into the tumor. In addition, growth factors and other angiogenic stimulators bound to the extracellular matrix can be released upon matrix proteolysis. The vascular cell receptors for bFGF and VEGF are specific transmembrane receptors that become phosphorylated on tyrosine residues upon activation. These events initiate signal transduction cascades, such as the MAP-kinase pathways, which in turn promote gene transcriptional events.

Programmed Cell Death, edited by Shi *et al.*
Plenum Press, New York, 1997

These angiogenic signals trigger the proliferation/invasion phase of angiogenesis, characterized by vascular cell division and secretion of both extracellular matrix (ECM) components and proteolytic enzymes which serve to remodel the extracellular microenvironment. This is accompanied by changes in the adhesive properties of the vascular cells, enabling them to respond to and interact with the remodeled ECM which serves to ensure continued cell proliferation and invasion.

Eventually, the vascular sprouts begins to mature as cell differentiation and lumen formation ensues. The new vessel secretes basement membrane components which serve to induce and maintain the endothelial cells in a differentiated and quiescent state.

2. INTEGRIN αVβ3 IN ANGIOGENESIS

Integrins are a family of heterodimeric plasma membrane receptors that mediates cell-ECM interactions and in some cases cell-cell adhesion (Hynes (1992)). Integrin function during blood vessel formation *in vivo* has best been studied for αVβ3, which is not detectable in quiescent vessels but become expressed at high levels after stimulus by angiogenic growth factors or tumors (Brooks et al (1994a)). Subsequent disruption of αVβ3-ligation by antibody or peptide antagonists of αVβ3 prevents blood vessel formation in the chick CAM, quail embryo, rabbit cornea, mouse retina and in human skin transplanted on SCID mice (Brooks et al (1994a), Brooks et al (1994b), Brooks et al (1995), Drake et al (1995), Friedlander et al (1995), Hammes et al (1996)). Importantly, this blockage of angiogenesis also results in a regression of human tumors in the chick embryo and inhibits human breast cancer growth by blocking human angiogenesis in skin grafts upon SCID mice (Brooks et al (1994b), Brooks et al (1995)). Upon further evaluation of the underlying mechanism, we found that proliferating angiogenic vessels that received anti-αVβ3 become apoptotic and that anti-αVβ3 blocked proliferation by inducing these proliferative endothelial cells to undergo apoptosis (Brooks et al (1994b)). We also demonstrate that angiogenic blood vessels, which are characterized by high αVβ3 expression, are those that become apoptotic upon treatment with αVβ3 antagonists (Brooks et al (1994b)). This suggests that integrin αVβ3 has a unique function during angiogenesis, i.e. to provide specific signals that are required for appropriate vascular cell proliferation. Integrin αVβ3 can interact with a number of ECM components such as vitronectin, osteopontin, fibronectin and denatured collagen. In addition, we recently found that integrin αVβ3 can also directly bind to matrix metalloproteinase-2 (MMP-2), thereby localizing matrix degradation capacity to the invasive/migratory site of vascular cells during angiogenesis (Brooks et al (1996)). This is another feature of integrin αVβ3 that might be important during angiogenesis, since collagen proteolysis by matrix metalloproteinases has been found to be crucial for blood vessel formation (Takigawa et al (1996), Moses et al (1990), Johnson et al (1994)). These observations may also shed light on how integrins and proteases function in a coordinate manner to promote cell invasion during angiogenesis.

However, integrin αVβ3 is not a pre-requisite for blood vessel development, since patients suffering from Glanzmann thrombobasthenia have apparently normal blood vessels, albeit a lacking integrin β3 subunit (Coller et al (1990)). Their capacity to develop blood vessels in the absence of αVβ3 might be explained by the fact that integrin αVβ5 can be used under certain conditions. Integrin αVβ3 is functional in angiogenesis stimulated by bFGF or TNF-α, whereas angiogenesis stimulated by VEGF or TGF-α depends on integrin αVβ5 (Friedlander et al (1995)).

3. CELL ADHESION AND APOPTOSIS

Integrins not only participate in cell adhesion and cytoskeletal re-organization, events necessary for cell migration, but they also mediate transmembrane signaling from the ECM into the cell Clark & Brugge (1995), Yamada & Miyamoto (1995)). Integrin-induced signals include H^+ exchange, Ca^{2+} influx, tyrosine-, serine- and threonine-phosphorylation events as well as altered phosphoinositide metabolism. In many cases, these signals have been linked to the regulation of gene expression and contribute to mechanisms such as proliferation, differentiation and cell survival. It has also been demonstrated that integrin-mediated cell adhesion can prevent apoptosis (Meredith et al (1993), Montgomery et al (1994)).

Considering the tight regulation of adhesion dependent proliferation of most non-transformed cells, a role for the integrin-ECM interactions in regulating apoptosis is well suited. However, studies in distinct cell systems under different conditions propose a role for various integrins as mediators of cell survival. Thus, the integrin required for cell survival and the survival signals generated might be cell type and condition specific. This may in part be explained by the distinct roles of different matrix components. Provisional matrices, including fibronectin, vitronectin and fibrinogen typically potentiate cell proliferation, whereas the basement membrane promotes cell cycle withdrawal and differentiation (Adams & Watt (1993), Lin & Bissell (1993)). Therefore, attachment to different ECM components is mediated through specific integrins, that can generate distinct signals that contribute to cell survival in multiple ways. To this end, Boudreau and colleagues recently demonstrated a relationship between the type of cell-ECM-interactions that occur and the specific signals that regulate apoptosis in mammary epithelial cells (Boudreau et al (1996)). Differentiation of mammary epithelial cells was induced by interaction with a three-dimensional basement membrane type matrix and accompanied by down-regulation of c-myc and cyclin D1, which normally promote proliferation. Once these cells became differentiated and quiescent, re-expression of c-myc resulted in apoptosis. In contrast, over-expression of c-myc in proliferating mammary cells was not accompanied by apoptosis (Boudreau et al (1996)). However, overexpression of p21[WAF-1/CIP-1], a cell cycle arrest molecule that is highly expressed during differentiation, lead to apoptosis when overexpressed in the proliferating mammary cells. These findings couple regulation of cell survival by the ECM intimately to cell cycle control, and suggest that introduction of signals which are in conflict with those mediated through the integrins may result in apoptosis.

4. INTEGRIN αVβ3 AND VASCULAR CELL SURVIVAL

Primary endothelial cells are anchorage-dependent and undergo apoptosis when denied integrin-mediated attachment (Meredith et al (1993), Re et al (1994)). These studies indicate that integrins might regulate endothelial cell survival *in vitro*. However, they do not adress integrin specificity or whether these events take place in complex biological mechanisms *in vivo* such as during angiogenesis. To this end, we found that specific signals mediated through integrin αVβ3 were required during angiogenesis *in vivo*, as blocking αVβ3 resulted in proliferating vascular cells becoming apoptotic (Brooks et al (1994a), Brooks et al (1994b), Brooks et al (1995)). Antagonists of αVβ3 systemically administered during angiogenesis in the chick CAM also activated endothelial cell p53 as measured by p53 dependent DNA binding activity, whereas antibodies to ß1 had no effect on p53 activity or apoptosis (Strömblad et al (1996), Brooks et al (1994b)). This induction

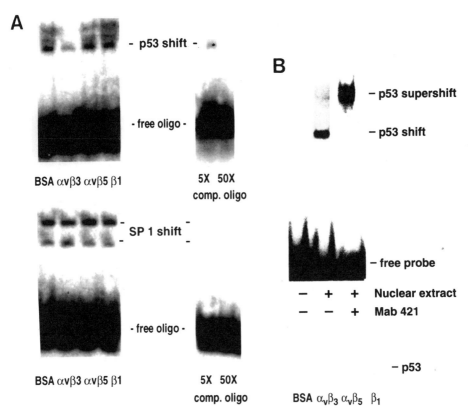

Figure 1. Ligation of endothelial cell integrin αvβ3 inhibits activation of p53. Human umbilical vein endothelial cells (HUVECs) were plated under serum free conditions on immobilized monoclonal antibodies LM 609 (anti-αvβ3), P4C10 (anti-β1), or PIF6 (anti-αvβ5) used as integrin agonists as previously described (10) for 4 h. Alternatively, cells were maintained in suspension for 4 h by preventing their attachment to culture dishes with 1% heat denatured bovine serum albumin (BSA). A. Nuclear extracts prepared from HUVECs attached as above were analyzed for p53 or SP1 DNA binding activity by electrophoretic mobility shift assay (EMSA) (left panel). Incubation in the presence of 5 or 50-fold excess unlabelled oligonucleotide in right panel. EMSAs using a mutant p53 recognition site did not reveal any mobility shift in these nuclear extracts (data not shown) B. Upper panel; Nuclear extract prepared from HUVECs that were prevented from adhering with heat-denatured BSA was analyzed in a super shift gel assay after preincubation in the presence or absence of anti-p53 Mab 421 (Oncogene Science, Cambridge, MA) at 4°C for 20 min. Lower panel; Lysates (15 ug) from HUVECs attached to various immobilized anti-integrin antibodies were analyzed for p53 protein levels by Western blot analysis using a polyclonal anti-p53 antibody (ab-7; Oncogene Science, Cambridge, MA). Reproduction from The Journal of Clinical Investigation, 1996, 98, 426–433, by copyright permission of The American Society for Clinical Investigation.

of p53 activity was also accompanied by an increased expression of the p53-inducible cell cycle inhibitor p21$^{WAF-1/CIP-1}$.

Using immobilized anti-integrin antibodies, functioning as agonists, we found that ligation of human endothelial cell αvβ3, but not of other integrins, was sufficient to suppress p53 activity without effecting its expression levels and to decrease p21$^{WAF-1/CIP-1}$ protein levels (Figures 1–2). These findings thus demonstrate that the ligation state of integrin αvβ3 directly regulates the cell cycle inhibitors p53 and p21$^{WAF-1/CIP-1}$ (Strömblad et al (1996)).

Furthermore, ligation of αvβ3 also enhances bcl-2 expression while decreasing that of bax (Figure 3). The resulting increase in the bcl-2/bax ratio has been previously shown

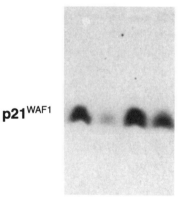

Figure 2. Ligation of endothelial cell integrin αvβ3 inhibits p21[WAF-1/CIP-1] expression. Lysates (15 μg protein) from cells attached as above were also analyzed for p21[WAF-1/CIP-1] protein levels by Western blot analysis using a polyclonal anti-p21[WAF-1/CIP-1] antibody (ab-5, Oncogene Science). Reproduction from The Journal of Clinical Investigation, 1996, 98, 426–433, by copyright permission of The American Society for Clinical Investigation.

to promote cell survival (White (1996)). In conclusion, ligation of endothelial cell αvβ3 acts to suppress apoptosis and conflicting growth arrest signals, thereby facilitating the proliferation and maturation of new blood vessels during angiogenesis.

It is still unclear whether activation of the cell cycle inhibitors p53 and p21[WAF-1/CIP-1] by αvβ3 antagonists is responsible for the blocking of angiogenesis. However, integrin αvβ3 mediates adhesion to proteins that typically comprise a provisional matrix, such as proteolyzed collagen, vitronectin, fibronectin and fibrinogen. As these matrix components promote proliferation, it might be expected that ligation of αvβ3 would in turn suppress the activity of molecules such as p53 or p21[WAF-1/CIP-1], which otherwise may inhibit cell cy-

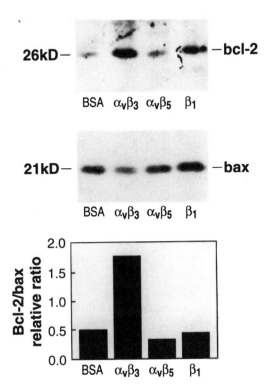

Figure 3. Ligation of endothelial cell integrin αvβ3 increases the bcl-2/bax ratio. HUVECs were plated under serum free conditions on immobilized Mabs LM 609 (anti-αvβ3), P4C10 (anti-β1), or PIF6 (anti-αvβ5) or were denied attachment on a BSA-coated surface for 4 h. Cells were then harvested and cell lysates (30 Fg) were analyzed for bcl-2 and bax protein levels by Western blotting using polyclonal anti-bcl-2 or anti-bax antibodies (sc 492 or sc 493; Santa Cruz Biotech., Santa Cruz, CA). The bars represent the relative bcl-2/bax ratio based upon quantities as measured by densitometry analysis of autoradiographs and do not represent an absolute molar ratio. Reproduction from The Journal of Clinical Investigation, 1996, 98, 426–433, by copyright permission of The American Society for Clinical Investigation.

cle progression and induce apoptosis. This suggests a model where integrin-extracellular matrix interactions are necessary to coordinate the activity of both cell cycle promotors and inhibitors. To prevent apoptosis during the proliferative phase of angiogenesis, interaction between αVβ3 and the provisional ECM would suppress conflicting growth arrest signals, such as p53 and p21[WAF-1/CIP-1] (Strömblad et al (1996)) (Figures 1–2). However, to promote survival during differentiation, the basement membrane would coordinately down-regulate cell cycle promotors when cell cycle inhibitors become activated. This may avoid a potential conflict that otherwise would lead to apoptosis, as has been observed when c-myc is introduced into mammary epithelial cells that are allowed to differentiate within a basement membrane-type matrix (Boudreau et al (1996)).

5. POTENTIAL LIGANDS FOR αVβ3 IN ANGIOGENESIS

Integrin αVβ3 is a promiscous receptor in that it can recognize a variety of RGD (Arg-Gly-Asp) containing extracellular matrix proteins, such as vitronectin, osteopontin, fibrinogen, fibronectin, denatured or proteolyzed collagen and thrombospondin. We have not yet been able to identify the physiological ligand(s) for integrin αVβ3 during angiogenesis. However, we have performed preliminary studies on the effect on some of these ligands on endothelial cell survival *in vitro*. We found that endothelial cells that are denied attachment by blocking with heat denatured BSA become apoptotic to a high degree within 4–5 h (Figure 4). A similar induction of apoptosis was observed when the human endothelial cells were plated on vitronectin, laminin, fibrinogen or native collagen type I, whereas cells attached to either heat denatured collagen type I or fibronectin survived to a much higher extend (Figure 4).

The fact that denatured but not native collagen type I can protect endothelial cells from apoptosis is intriguing, since our studies indicate that collagen degradation takes

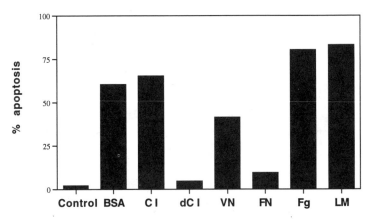

Figure 4. Endothelial cell survival upon different extracellular matrix proteins. HUVECs were plated under serum-free conditions on pre-coated dishes with different extracellular matrix proteins (25ug/ml) or kept in suspension by heat denatured BSA (1%). After 5 h, cells were harvested and stained for fragmented DNA using Apop Tag (Oncor, Gaithsburg, MD) followed by FACS Scan analysis (ref Brooks 1994b). The graph shows the procentage of positively staining cells among attached cells, except when all cells were maintained in suspension by BSA blocking. Control - HUVE cells harvested directly from serum-containing culturing conditions. BSA - Cells maintained in suspension. CI - Collagen type I. dCI - Heat denatured collagen type I. VN - Vitronectin. FN - Fibronectin. Fg - Fibrinogen. LM - Laminin.

place around angiogenic blood vessels expressing integrin αVβ3 (S. Strömblad, preliminary results). Also, the prescence and activity of collagenases has been found to be critical for angiogenesis (Takigawa et al (1996), Moses et al (1990), Johnson et al (1994)). Attachment to native collagen type I is mediated by β1-integrins, in particular αVβ3. However, when collagen is proteolyzed or heat denatured, cryptic RGD-containing sites get exposed, which can be utilized by integrin αVβ3 for attachment and migration (Davis (1993), Mongomery et al (1994), Pfaff et al (1994)). Thus, proteolyzed collagen type I is a putative ligand for integrin αVβ3 during angiogenesis that also qualifies in its ability to protect vascular cells from apoptosis.

6. CONCLUSIONS

We have found that integrin αVβ3 is preferentially expressed on newly forming blood vessels and that antagonists of integrin αVβ3 blocks angiogenesis by inducing apoptosis in proliferating vascular cells. This results in tumor regression or growth inhibition of various human tumors. In addition, our results provide a link between integrin αVβ3 ligation, p53 activity, expression of the cell cycle inhibitor $p21^{WAF-1/CIP-1}$, and vascular cell survival. Thus, during angiogenesis, ligation of endothelial cell αVβ3 is required for the suppression of apoptosis and of conflicting growth arrest signals, thereby facilitating the proliferation and maturation of new blood vessels.

Furthermore, we indicate that proteolyzed collagen type I might be a ligand for endothelial cell integrin αVβ3 during angiogenesis, since cryptic RGD-site(s) in collagen type I that are exposed upon denaturation are capable of mediating endothelial cell survival.

REFERENCES

Adams, J.C. & Watt, F.M. (1993). Regulation of development and differentiation by the extracellular matrix. Development 117, 1183–1198.

Boudreau, N., Werb, Z. & Bissell, M.J. (1996). Suppression of apoptosis by basement membrane requires three-dimensional tissue organization and withdrawal from the cell cycle. Proc. Natl. Acad. Sci. U.S.A. 93, 3509–3513.

Brooks, P.C., Clark, R.A. & Cheresh, D.A. (1994a). Requirement of vascular integrin αVβ3 for angiogenesis. Science 264, 569–571.

Brooks, P.C. et al (1994b). Integrin αvβ3 antagonists promote tumor regression by inducing apoptosis of angiogenic blood vessels. Cell 79, 1157–1164.

Brooks, P.C., Strömblad, S., Klemke, R., Visscher, D., Sarkar, F.H. & Cheresh, D.A. (1995). Antiintegrin αVβ3 blocks human breats cancer growth and angiogenesis in human skin. J. Clin. Invest. 96, 1815–1822.

Brooks, P.C., Strömblad, S., Sanders, L.C., von Schalscha, T., Aimes, R.T., Stetler-Stevenson, W.G., Quigley, J.P. & Cheresh, D.A.(1996). Localization of matrix metalloproteinase MMP-2 to the surface of invasive cells by interaction with integrin αVβ3. Cell 85, 683–693.

Clark, E.A. & Brugge, J.S. (1995). Integrins and signal transduction pathways: the road taken. Science 268, 233–239.

Coller, B.S., Cheresh, D.A., Asch, E. & Seligsohn, U. (1991). Platelet vitronectin receptor expression differentiates Iraqui-Jewish from Arab patients with Glanzmann thrombobastenia in Israel. Blood 77, 75–83.

Drake, J., Cheresh, D.A. & Little, C.D. (1995). An antagonist of integrin αVβ3 prevents maturation of blood vessels during embyonic neovascularization. J. Cell Sci. 108, 2655–2661.

Folkman, J. (1995). Angiogenesis in cancer, vascular rheumatoid and other disease. Nature Med. 1, 27–30.

Friedlander, M., Brooks, P.C., Shaffner, R.W., Kincaid, C.M., Varner, J. & Cheresh, D.A. (1995). Definition of two angiogenic pathways by distinct αv integrins. Science 270, 1500–1502.

Hammes, H.-P., Brownlee, M., Jonczyk, A., Sutter, A. & Preissner, K.T. (1996). Nature Med. 2, 529–533.

Hynes, R.O. (1992). Integrins: versatility, modulation, and signaling in cell adhesion. Cell 69, 11–25.

Johnson, M.D., Kim, H.-R.C., Chesler, L., Tsao-Wu, G., Bouck, N. & Polverini, P.J. (1994). Inhibition of angiogenesis by tissue inhibitor of metalloproteinase. J. Cell. Physiol. 160, 194–202.

Lin, C.Q. & Bissell, M.J. (1993). Multi-facetted regulation of cell differentiation by extracellular matrix. FASEB J. 7, 737–743.

Meredith, J.E., Fazeli, B. & Schwartz, M.A. (1993). The extracellular matrix as a survival factor. Mol. Biol. Cell 4, 953–961.

Montgomery, A.M.P., Reisfeld, R.A. & Cheresh, D.A. (1994). Integrin $\alpha V \beta 3$ rescues melanoma cells from apoptosis in three-dimensional dermal collagen. Proc. Natl. Acad. Sci. U.S.A. 91, 8856–8860.

Moses, M.A., Sudhalter, J., Lanfer, R. (1990). Identification of an inhibitor of neovascularization from cartilage. Science 248, 1408–1410.

Pfaff, M., McLane, M.A., Beviglia, L., Niewiarowski, S. & Timpl, R. (1994). Comparison of disintegrins with limited variation in the RGD loop in their binding to purified integrins alpha IIb beta 3, alpha v beta 3 and alpha 5 beta 1 and in cell adhesion inhibition. Cell Adhes. Commun. 2, 491–501.

Re, F. et al., & Colotta, F. (1994). Inhibition of anchorage-dependent cell spreading triggers apoptosis in cultured human endothelial cells. J. Cell Biol. 127, 537–546.

Strömblad, S., Becker, J.C., Yebra, M., Brooks, P.C. & Cheresh, D.A. (1996). Suppression of p53 activity and p21[WAF-1/CIP-1] expression by vascular cell integrin $\alpha V \beta 3$ during angiogenesis. J. Clin. Invest. 98, 426–433.

Takigawa, M., Yukimitsu, Y., Suzuki, F., Kishi, J-I., Yamashita, K. & Hayakawa, T. (1990). Induction of angiogenesis in chick yolk-sac membrane by polyamines and its inhibition by tissue inhibitors of metalloproteinases (TIMP and TIMP-2). Biochem. Biophys. Res. Commun. 171, 1264–1271.

Weidner, N. (1995). Intratumor microvessel density as a prognostic factor in cancer. Am. J. Pathol. 147, 9–19.

White, E. (1996). Life, death, and the pursuit of apoptosis. Genes & Dev. 10, 1–15.

Yamada, K.M. & Miyamoto, S. (1995). Integrin transmembrane signaling and cytoskeletal control. Curr. Opin. Cell Biol. 7, 681–689.

5

THE ROLE OF GRANZYME B IN CYTOTOXIC T LYMPHOCYTE ASSISTED SUICIDE

R. Chris Bleackley

Department of Biochemistry
University of Alberta
Edmonton, Alberta T6G 2H7, Canada.

1. INTRODUCTION

Two branches have evolved in the immune system to recognize foreign molecules and thus protect against pathogens. In humoral immunity, antibodies are produced that bind to these specific structures and hence stimulate their removal by phagocytosis or complement mediated lysis. T lymphocytes have on their surface a receptor that, like antibodies, binds to foreign peptides presented in the context of the major histocompatibility proteins. As a result the effectors of cell mediated immunity, the cytotoxic T lymphocytes, become activated and can then recognize and destroy any antigen expressing pathogenic cell. This branch of the immune system is particularly involved in the eradication of virus infected cells, but also may play a role in lysis of tumors, transplanted cells and in the destruction of normal host cells in autoimmune disorders.

Complement mediated cell lysis involves a cascade of proteolytic activities that culminates in the activation of a membraneolytic complex. We will see here that CTL induced killing also involves related enzymes but culminates in a much more subtle form of cell death. The mechanism involves the activation of a preprogrammed death pathway within the targets that exists in the majority of cells. Indeed it is perhaps more appropriate to think of this pathway as a cytotoxic lymphocyte assisted suicide, or CLASsier form of death.

2. CTL-SPECIFIC GENES

In our initial search for molecules that may be important in the killing process, we hypothesized that potential effector proteins would be expressed uniquely in activated cytotoxic T lymphocytes. We therefore screened a cDNA library from a CTL line with probes generated from the mRNA of activated and resting T cells, EL4 (as a helper T cell equivalent) and brain cells (Lobe *et al.*, 1986a). Only clones which hybridized specifically with the first probe were considered for further analysis.

Obviously this kind of approach can lead to an embarrassment of riches and a strategy must be developed to focus on genes that are more likely to encode important molecules for the process in question. We were struck by the fact that two of our cDNAs were predicted to encode serine proteases (Lobe *et al.*, 1986b). These enzymes recognize sequences within proteins and catalyze the hydrolysis of specific peptide bonds. As a direct consequence of this protein cleavage a series of events are set in motion that results in a biological effect. This could be as simple as activation of a growth factor but these enzymes are of vital importance in a wide variety of biological processes and are critical in blood clotting and complement mediated lysis. With a defined and potentially important enzymatic activity we decided to pursue these cytotoxic cell proteases (CCP) further.

2.1. A Multigene Family

Before proceeding I should point out that these proteases have been christened by numerous investigators and, in addition to CCP, are referred to variously as CTLA (cytotoxic T lymphocyte antigen), SE (serine esterase), TSP (T cell serine protease) and CSP (cytotoxic serine protease), but the name that has gained wide acceptance is granzyme (granule associated enzyme). The granzymes form part of a multigene family (grB to G) that are clustered, close to the α chain of the T cell receptor gene, on chromosome 14 (Crosby *et al.*, 1990). Other serine proteases including some expressed in mast cells and neutrophils are also found in this same locus. In addition there is another granzyme, referred to as A or HF (Hannukah Factor) that is encoded in at a separate locus (Gershenfeld & Weissman, 1986).

The genomic organization of all the chromosome 14 granzymes is remarkably conserved, with almost complete retention of intron splice sites (Bleackley *et al.*, 1988a; Prendergast *et al.*, 1991). In addition the proteins themselves are very similar at the primary sequence level (Bleackley *et al.*, 1988b). All are predicted to be synthesized with a hydrophobic leader sequence that is cleaved off as the nascent protein is translocated into the endoplasmic reticulum. They exist in an inactive form by virtue of just two extra aminoacids at their amino termini (Caputo *et al.*, 1993). These so called zymogens are likely converted to the active enzymes as the proteins are processed in the golgi, just prior to packaging in the secretory granules (see later).

2.2. Granzyme Expression Correlates with Cytolytic Activity

One of the key findings that convinced us that the granzymes were involved in killing was the close correlation between expression and lytic activity (Lobe *et al.*, 1986a; 1986b). Initially we believed that granzyme gene transcription was coordinately induced and preceded cytolytic activation under a variety of activation conditions. Later when we developed PCR methods that could distinguish between family members, it became clear that the granzymes were not all expressed together. More likely, their transcription is independently regulated, and this may relate to their specific functions (Prendergast *et al.*, 1992). Isolation and characterization of *cis*-acting regulatory elements that control granzyme transcription also points to differential expression (Lobe *et al.*, 1989; Fregeau & Bleackley, 1991; Babichuk *et al.*, 1996).

Granzyme B (CCP1) transcription correlates most closely with the acquisition of cytotoxicity by the T cell (Prendergast *et al.*, 1992). Whether we activate cells with antigen, mitogen, anti CD3 or IL2, the peak of grB mRNA always precedes the maximum level of cytotoxicity. In addition to numerous *in vitro* studies, this correlation also hold true *in vivo* in both mouse (Mueller *et al.*, 1988) and human (Lipman *et al.*, 1992). Thus in a mouse model of cardiac transplant rejection there is a close correlation between infiltration by the transplanted tissue by granzyme positive T lymphocytes and rejection. Similarly in the human experiments expression of grB, assayed by PCR, coincided with pathological diagnosis of rejection in a series of kidney transplant recipients. The studies in human are particularly interesting as they indicate that expression of grB may be a useful marker of rejection that could be of clinical relevance for patients who have received transplants.

2.3. Granzyme B Is an Aspase

The pattern of expression of grB is tantalizing but the molecule itself is of inherent biochemical interest. With just the primary sequence of grB in hand my colleagues Dr. M.N.G. James and M. Murphy were able to predict a 3D model of the structure. This was based on the coordinates determined for the related mast cell protease II. Overall the structures appeared to be similar but grB had an arginine prominently inserted into the substrate binding pocket. This lead us to propose that grB would hydrolyze proteins at acidic residues, most likely aspartic acid (Murphy *et al.*, 1988). An hypothesis that was quite bold, as no known eukaryotic serine protease had such a substrate specificity.

We decided to test this prediction using a molecular genetic approach. First we developed an heterologous expression system for grB in which COS cells were transfected with the grB cDNA. The extracts were tested for Aspase activity with a synthetic substrate AlaAlaAsp-thiobenzyl ester. The wild type gene gave us no activity, but, to our delight, when we programmed the system with a cDNA in which the putative zymogen dipeptide was removed we saw thiobenzyl ester being released from the substrate (Caputo *et al.*, 1993). These same extracts did not hydrolyze chymotrypsin or trypsin substrates but when we mutated the Arg, that was the focus of our attention, we converted the enzyme from an Aspase to a chymase (Caputo *et al.*, 1994). Thus the prediction of the Aspase activity as well as the theoretical basis for the prediction were proven to be correct. We will see later that this was of fundamental importance in defining a physiologically relevant substrate for grB.

2.4. Granzymes Induce Apoptosis in Targets

Another important indication of the importance of the granzymes was the fact that they were sequestered within the cytoplasm of cytotoxic cells in granules (Redmond *et al.*, 1987). In the granule-mediated mechanism of killing that had been proposed, potential effectors of cytolysis are delivered to the target cell by the CTL via a process of vectoral exocytosis (Henkart & Henkart, 1982). One of the granule-proteins, perforin, was indeed cytolytic in its own right and for a while was believed to be sufficient for death. However, it was then realized that perforin alone was capable only of inducing necrosis (Duke *et al.*, 1989). In contrast intact CTL, or indeed purified granules, were able to induce both membrane damage and DNA fragmentation within the target. This form of cell death, in which DNA fragmentation into oligonucleosomal fragments occurs, is often seen during development and is known as apoptosis. The proteins responsible for this DNA fragmenting ability (fragmentins) were ultimately purified by Greenberg *et al.*, and shown to be identical to the granzymes (Shi *et al.*, 1992). It is now believed that perforin acts to facilitate uptake of the granzymes into the target cell where they cleave their relevant substrates. This sets in motion a series of events which culminates in the death of the target by apoptosis.

2.5. Cytotoxicity Enters the Ice Age

Apoptosis was described over 25 years ago but was largely ignored until the oncogene bcl2 was shown to act by blocking this form of cell death. Since then there has been an incredible amount of effort expended in order to understand the molecular events involved. One of the most rewarding biological systems to study programmed cell death has been the nematode *C. elegans.* I had the good fortune to hear Bob Hortwitz describe the identification of a protein involved in nematode cell death named ced3. This molecule was related to the mammalian interleukin 1β converting enzymes (ICE) and required proteolytic activation in order to bring about apoptotic cell death (Yaun *et al.,* 1993). However, this was not just any proteolytic event as the sequence in question, in the ced3 protein, involved cleavage at an Asp residue. Of course this struck an immediate note for me as it was exactly the same specificity that we had defined for granzyme B.

Initially the only mammalian homologue to ced3 was ICE itself, but we were able to demonstrate that this was not a substrate for grB (Darmon *et al.,* 1994). However, in an amazingly short period of time the ICE-family grew and is still expanding. The most likely candidate for a mammalian ced3 was the protein CPP32/Apopain/Yama. This enzyme had been purified on the basis of its ability to cleave poly (ADP-ribose)polymerase (PARP), a hallmark event of *in vitro* apoptosis (Nicholson *et al.,* 1994).

Through the generosity of Don Nicholson (Merck Frosst, Montreal) we were able to show that grB could indeed cleave and activate CPP32 (Darmon *et al.,* 1995). The gene for CPP32 was transcribed and translated using an *in vitro* system. When the radioactive CPP32 protein was incubated with either purified granule grB or recombinant active enzyme, we saw clear evidence of conversion from p32 to p17. Activation was then confirmed by using PARP as a substrate for the CPP32 cleaved by grB. With the heterologous expression system, that we had developed for grB, we further demonstrated the absolute requirement for grB catalytic activity in the cleavage process. Finally we backed up the relevance of our *in vitro* studies with a western blot experiment that clearly established that CPP32 was cleaved in target cells under attack by whole CTL.

2.6. The grB/CPP32 Pathway Leads to DNA Fragmentation

In order to ascertain whether the *in vitro* cleavage of CPP32 by grB was linked to activation of CPP32 in CTL induced target cell killing, we generated CTL lines from grB knock out mice. When these were used as effectors we saw no evidence of CPP32 processing (Darmon *et al.,* 1996). These CTL can still induce membrane damage in targets but DNA fragmentation is significantly reduced. Thus we suppose that the grB defect and consequently the lack of CPP32 activation leads primarily to consequences at the level of the genome. In confirmation of this we further found that a "specific inhibitor" (DEVD-CHO) of CPP32 activity also has its major impact on DNA degradation (Darmon *et al.,* 1996). Although we cannot rule out the involvement of other CPP32-like enzymes, a similar inhibitor targetted to ICE (YVAD-CHO) was unable to suppress either read out of granzyme-mediated apoptosis.

Although it would appear that the grB-CPP32 pathway is involved in DNA fragmentation, or more likely lack of ability to repair this genomic damage, *in vitro* studies clearly implicate grB in other features characteristic of apoptosis including membrane damage. Thus we assume that there are other substrates for granzyme B that would explain this. In an attempt to identify these molecules we have developed a method that allows us to detect and, we hope, identify grB-binding proteins (Pinkoski *et al.,* 1996).

Cells are permeabilized after they have been fixed on microscope slides. After incubation with grB, the samples are then treated with anti grB antibody and finally a secondary detection antibody. As a negative control we have used zymogen grB produced in our recombinant expression system. Thus we can then localize the potential substrates, and other important interacting molecules by microscopy. We find some cytoplasmic binding but were struck by the nuclear localization, particularly in regions that appear to be perinucleolar. This method has now been adapted for immunoprecipitation using a catalytically inactive mutant of grB (to prevent proteolysis). A number of proteins can be precipitated from both the cytoplasm and nucleus of targets. Further characterization of these molecules is currently underway, but it is hoped that they will include not only substrates but also other important interacting proteins that may, for example, play a role in the control of intracellular transport of grB. Recent experiments indicate that grB localization in the target cell could be quite important. When cells are treated with grB alone uptake is delayed but build does occur in the cytoplasm. However if a second agent such as perforin is added, the grB appears in the nucleus of apoptotic cells.

Other ICE-family members have already been identified as *in vitro* substrates of grB (Gu *et al.*, 1996). It remains a significant challenge for future work to understand the *in vivo* relevance of, and possible cross-talk between each of these events.

3. GRANULE-INDEPENDENT KILLING

For some time it has been apparent that target cell lysis can be seen in the absence of "obvious" granules and can sometimes occur in the absence of calcium, a requirement for degranulation and perforin action. Our own studies in this area were stimulated by Gideon Berke who have developed a series of peritoneal exudate lymphocyte (PEL) hybridomas that killed but contained no granules. Using both a very sensitive PCR assay for granule-protein transcripts and immunocytochemistry we confirmed his assertion that these cells expressed neither perforin or granzymes (Helgason *et al.*, 1992). In contrast to the granule-mediated killing process, lysis by PEL was insensitive to calcium chelation and cyclosporin treatment. However, both pathways were apparently independent of protein synthesis in the targets. The important clue to this alternate mechanism came from the correlation of target cell sensitivity with the level of Fas-antigen expression on the surface (Garner *et al.*, 1994). Thus the PEL cells were able to lyze Fas$^+$ targets because they expressed the Fas-ligand after activation.

3.1. ICE-Enzymes in Fas-Mediated Lysis

At the time we became involved with experiments to investigate the role of ICE in granule-mediated killing, it seemed that the PEL cells would represent an ideal control for a protease-independent lytic mechanism. We were very surprised, however, to find that the ICE inhibitor YVAD aldehyde, although unable to interfere with granule-mediated killing was a potent inhibitor of Fas induced lysis. In contrast the CPP32 inhibitor, DEVD aldehyde, was without effect in the Fas system. Although in some cell types we do see some inhibition of PEL induced DNA fragmentation with this latter inhibitor (Darmon & Bleackley, 1996).

It would seem therefore that ICE-like enzymes are also involved in this alternate killing mechanism. Our evidence suggests that it is not ICE itself as other potent ICE inhibitors are not effective, and we have been unable to demonstrate interleukin 1β converting activity in cells programmed to die by PEL. There is, however, an activity present that can cleave the commonly used ICE activity peptide. We conclude that an ICE-enzyme, not CPP32 or ICE, plays a key role in the process of Fas-induced apoptosis. It will be very interesting to test whether the newly described Fas associated ICE enzyme (FLICE) has these same properties.

4. GRANZYMES A DECADE LATER

After working on these interesting enzymes for the past 10 years it is certainly grati-
fying to see them established as major effectors of CTL-induced killing. Granzyme B in
particular seems to play a pivotal role in inducing the target cell to turn on its own programmed cell death pathway.

My own view of the events involved in the induction of apoptosis within target cells
is presented in Figure 1. On the left hand side we have the pathway that has been the focus
of our attention for the last few years. In combination perforin and granzyme B act as a
target cell to cleave and activate CPP32. Although we have provided convincing evidence
that the action of grB on CPP32 is direct *in vitro*, and that undoubtedly the two proteases
are on the same pathway *in vivo*, it is still not proven that grB cleaves and activates CPP32
in vivo. This part of the pathway impinges primarily on DNA fragmentation, perhaps most
likely involving inactivation of DNA repair enzymes. However, it is clear from the grB
knock out CTL and from the results with CPP32 inhibitors that late DNA fragmentation
and membrane damage occur via a different route.

It is likely that the grB independent DNA fragmentation results from a combination
of Fas-mediated killing and via the involvement of other granzymes. I would say that the
majority of evidence points to grA as the major mediator in the latter. As indicated on the
right hand side of the figure we have, at present, very little knowledge of the molecules involved in this pathway.

In addition to DNA fragmentation a number of other events have been documented as
characteristic of apoptosis. Many of these involve destruction, or perhaps more correctly re-
organization, of membrane components. These proceed unabated in the presence of CPP32
inhibition. Perforin was originally isolated on the basis of its membrane perforating activity.
It is not clear if there is enough perforin delivered by CTL to bring about significant
amounts of membrane disruption. While the E:T ratios of 20:1 that are commonly used *in
vitro* are rare *in vivo*, this remains a possible pathway. Experimentally it has been shown
that sublytic doses of perforin can produce significant amounts of membrane damage when

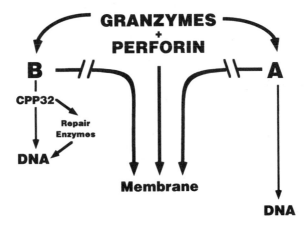

Figure 1. Pathways to granule mediated killing. Granzyme B through cleavage of CPP32 and related enzymes is a
key mediator of DNA damage. However, late DNA damage is likely also mediated through a grA branch. Perforin,
grA and grB likely all contribute to membrane events.

supplemented with granzymes. I believe it is more likely that the granzymes have substrates, that remain to be identified, that will impinge on these other apoptosis events.

I think that we now have an overview of the first stages of the pathway but much work is required to understand how the cell actually meets in demise. The roles of multiple granzymes in the CTL, and the presence of many ICE-enzymes in the targets is presently unclear. We do not have an explanation for the induction of either DNA fragmentation or membrane damage. The relationship between the ICE-enzymes and their downstream substrates remains to be elucidated. It is unclear whether internalization of grB is required to activate the process or whether targetting of the granzyme to a specific intracellular site is required. Indeed most experimental evidence indicates that a second signal is required to induce apoptosis.

Outstanding questions apart, our increased understanding of CTL-mediated killing has important implications for drug design and screening. The cleavage of CPP32 by grB is an important indicator of CTL activity and thus has the potential to be used as an impartial surrogate maker of CTL activity. It is already clear that one strategy that viruses employ to evade the immune system is the inhibition of apoptosis. Perhaps screening for compounds that inhibit or enhance the cleavage and activation of CPP32 by grB could have an impact on autoimmune or transplantation, and cancer or infectious diseases respectively. The expression of grB activity has already proven to be a key marker of CTL activation in the clinic, and has significant potential for screening new vaccine formulations and immunotherapeutics. We still don't understand the killing process in molecular detail, but the past decade has indeed been rewarding for both basic biomedical understanding and potential therapeutic intervention. The next ten years may see significant applications of this knowledge.

ACKNOWLEDGMENTS

I have been fortunate in the number of talented students, fellows and technologists who have chosen to work in my laboratory over the years. Without their experimental talents, hard work and intellectual curiosity we would have made little progress. The work was made possible by funding from the National Cancer Institute and the Medical Research Council of Canada. Scholarships and Fellowships from the Alberta Heritage Foundation for Medical Research, the National Cancer Institute, the Medical Research Council, the National Science and Engineering Council and the University of Alberta. The author is a Medical Scientist of the AHFMR. As always Mae Wylie provided me with superb secretarial support for this manuscript.

REFERENCES

Babichuk, C.K., B.L. Duggan & R.C. Bleackley. *In vivo* regulation of murine granzyme B gene transcription in activated primary T cells. *J. Biol. Chem.* **271**:16485–16493 (1996).

Bleackley, R.C., C.G. Lobe, B. Duggan, N. Ehrman, C. Fregeau, M. Meier, M. Letellier, C. Havele, J. Shaw & V. Paetkau. The isolation and characterization of a family of serine protease genes expressed inactivated cytotoxic T lymphocytes. *Immunological Reviews* **103**:5–19 (1988a).

Bleackley, R.C., B. Duggan, N. Ehrman & C.G. Lobe. Isolation of two cDNA sequences which encode cytotoxic cell proteases. *FEBS Lett.* **234**:153–159 (1988b).

Caputo, A., M.N.G. James, J.C. Powers, D. Hudig & R.C. Bleackley. Conversion of the substrate specificity of mouse proteinase granzyme B. *Nature Structural Biology* **1**:364–367 (1994).

Caputo, A., R.S. Garner, U. Winkler, D. Hudig & R.C. Bleackley. Activation of recombinant murine cytotoxic cell proteinase-1 requires deletion of an amino-terminal dipeptide. *J. Biol. Chem.* **268**:17672–17675 (1993).

Crosby, J.L., R.C. Bleackley & J.H. Nadeau. A complex of serine protease genes expressed preferentially in cytotoxic T-lymphocytes is closely linked to the T-cell receptor å- and _-chain genes on mouse chromosome 14.
Genomics **6**:252–259 (1990).

Darmon, A. & R.C. Bleackley. An ICE-like protease is a key component of fas-mediated apoptosis. *J. Biol. Chem.*
00:000–000 (1996). (in press)

Darmon, A.J., D.W. Nicholson & R.C. Bleackley. Activation of the apoptotic protease CPP32 by cytotoxic T-cell-
derived granzyme B. *Nature* **377**:446–448 (1995).

Darmon, A.J., N. Ehrman, A. Caputo, J. Fujinaga & R.C. Bleackley. The cytotoxic T cell proteinase granzyme B
does not activate interleukin-1_-converting enzyme. *J. Biol. Chem.* **269**:32043–32046 (1994).

Darmon, A.J., T.J. Ley, D.W. Nicholson & R.C. Bleackley. Cleavage of CPP32 by granzyme B represents a nonre-
dundant role for granzyme B in the induction of target cell DNA fragmentation. *J. Biol. Chem.* **00**:000–000
(1996). (in press)

Duke, R.C., P.M. Persechini, S. Chang, C.-C. Liu, J.J. Cohen & J.D.-E. Young. Purified perforin induces target cell
lysis but not DNA fragmentation. *J. Exp. Med.*, **170**:1451–1456 (1989).

Frégeau, C.J. & R.C. Bleackley. Transcription of two cytotoxic cell protease genes is under the control of different
regulatory elements. *Nucleic Acids Research* **19**:5583–5590 (1991).

Garner, R., C.D. Helgason, E.A. Atkinson, M.J. Pinkoski, H.L. Ostergaard, O. Sorensen, A. Fu, P.H. Lapchak, A.
Rabinovitch, J.E. McElhaney, G. Berke & R.C. Bleackley. Characterization of a granule-independent lytic
mechanism used by CTL hybridomas. *J. Immunol.* **153**:5413–5421 (1994).

Gershenfeld, H.K. & I.L. Weissman. Cloning of a cDNA for a T cell-specific serine protease from a cytotoxic T
lymphocyte. *Science* **232**:854–858 (1986).

Gu, Y., C. Sarnecki, M.A. Fleming, J.A. Lippke, R.C. Bleackley & M.S.-S. Su. Processing and activation of CMH-
1 by granzyme B. *J. Biol. Chem.* **271**:10816–10820 (1996).

Helgason, C.D., J.A. Prendergast, G. Berke & R.C. Bleackley. Peritoneal exudate lymphocyte and mixed lymphocyte culture hybridomas are cytolytic in the absence of cytotoxic cell protease and perforin. *Eur. J. Immunol.* **22**:3187–3190 (1992).

Henkart, M.P. & P.A. Henkart. Lymphocyte mediated cytolysis as a secretory phenomenon. *Adv. Exp. Med. Biol.*
146:227–247 (1982).

Lipman, M.L., A.C. Stevens, R.C. Bleackley, H. Helderman, T.R. McCune, W.H. Harmon, M.E. Shapiro, S. Rosen
& T.B. Strom. The strong correlation of cytotoxic T lymphocyte-specific serine protease gene transcripts
with renal allograft rejection. *Transplantation* **53**:73–79 (1992).

Lobe, C.G., C. Havele & R.C. Bleackley. Cloning of two genes which are specifically expressed in activated cytotoxic T lymphocytes. *Proc. Natl. Acad. Sci. USA* **83**:1448–1452 (1986a).

Lobe, C.G., B. Finlay, W. Paranchych, V.H. Paetkau & R.C. Bleackley. Two cytotoxic T lymphocyte-specific
genes encode unique serine proteases. *Science* **232**:858–861 (1986b).

Lobe, C.G., J. Shaw, C. Fregeau, B. Duggan, M. Meier, A. Brewer, C. Upton, G. McFadden, R.K. Patient,
V.H. Paetkau & R.C. Bleackley. Transcriptional regulation of two cytotoxic T lymphocyte-specific serine
protease genes. *Nucleic Acids Research* **17**:5765–5779 (1989).

Mueller, C., H.K. Gershenfeld, C.G. Lobe, C.Y. Okada, R.C. Bleackley & I.L. Weissman. Expression of two serine
esterase genes during an allograft rejection in the mouse. *Transplantation Proc.* (1988).

Murphy, M.E.P., J. Moult, R.C. Bleackley, I.L. Weissman & M.N.G. James. Comparative molecular model build-
ing of two serine proteinases from cytotoxic T lymphocytes. *Proteins: Structure, Function and Genetics*
4:190–204 (1988).

Pinkoski, M.J., U. Winkler, D. Hudig & R.C. Bleackley. Binding of granzyme B in the nucleus of target cells: Recognition of an 80 kDa protein. *J. Biol. Chem.* **271**:10225–10229 (1996).

Prendergast, J.A., C.D. Helgason & R.C. Bleackley. Quantitative polymerase chain reaction analysis of cytotoxic
cell proteinase gene transcripts in T cells - pattern of expression is dependent on the nature of the stimulus.
J. Biol. Chem. **267**:5090–5095 (1992).

Prendergast, J.A., M. Pinkoski, A. Wolfenden & R.C. Bleackley. Structure and evolution of the cytotoxic cell proteinase genes CCP3, CCP4 and CCP5. *J. Mol. Biol.* **220**:867–875 (1991).

Redmond, M.J., M. Letellier, J.M.R. Parker, C. Lobe, C. Havele, V. Paetkau & R.C. Bleackley. A serine protease
(CCP1) is sequestered in the cytoplasmic granules of cytotoxic T lymphocytes. *J. Immunol.*
139:3184–3188 (1987).

Shi, L., C.-M. Kam, J.C. Powers, R. Aebersold & A.H. Greenberg. Purification of three cytotoxic lymphocyte
granule serine proteases that induce apoptosis through distinct substrate and target cell interactions. *J. Exp.
Med.* **176**:1521–1529 (1992).

Yuan, J., S. Shaham, S. Ledoux, H.M. Ellis & H.R. Horvitz. The *C. elegans* cell death gene ced-3 encodes a protein similar to mammalian interleukin-1B-converting enzyme. *Cell* **75**:641–652 (1993).

6

CASPASE INHIBITORS AS MOLECULAR PROBES OF CELL DEATH

Apurva Sarin,[1] Klaus Ebnet,[1] Charles M. Zacharchuk,[2] and Pierre A. Henkart[1]

[1]Experimental Immunology Branch
[2]Laboratory of Immune Cell Biology
National Cancer Institute
National Institutes of Health
Bethesda, Maryland 20892

1. INTRODUCTION

Recognition of the importance of programmed cell death in basic biological processes as well as its relevance to clinical problems has led to a surge of research interest in this area in recent years. The appealing idea behind much of this research is that an endogenous death pathway is present in most cells which can be triggered to kill cells by physiological signals or when they encounter some types of stressful conditions. In spite of the rapid progress made in some respects, fundamental issues in this field remain unresolved, and the major challenge is still to molecularly characterize this death pathway and understand the controls which lead to its activation. Recent research in a number of laboratories has identified a family of enzymes which at very least now provides a testable candidate for a central player in such physiological death pathways. Specific inhibitors of these enzymes are newly available tools which can be used to test for their participation in particular physiological and experimental cell death systems (Henkart, 1996). In this paper we will present examples illustrating their use from our own studies on lymphoid cells. For reasons which will become clear, we believe that a functional death criterion based on inhibiting enzyme function provides a more significant characterization of cell death than the commonly used apoptotic death phenotypes.

2. THE ROLE OF CASPASES IN CELL DEATH

The critical event in the identification of the molecular family central to programmed cell death came from studies of the nematode *C. elegans* death gene *ced-3,* which controls the normal developmental cell deaths during embryonic development (Hengartner and Horvitz, 1994). The protein sequence encoded by this gene was unexpectedly found to be homologous to the mammalian protease interleukin-1-β converting enzyme (ICE), and in the last several years about ten related human enzymes have been characterized by several groups. This unique family of proteases was recently named caspases after their shared properties of being cysteine proteases which cleave substrates after aspartic acid residues (Alnemri et al., 1996). There are at least two subfamilies of caspases based on sequence homology as well as the fine specificity of substrate cleavage. The ICE-like subfamily prefers a hydrophobic residue at the substrate P4 position, while the CPP-32-like subfamily prefers an aspartate at P4. It has become clear that the various caspase family proteins are widely expressed in the cytoplasm of many cell types. They are present in an inactive precursor form, and cell death stimuli lead to their proteolytic processing and consequent enzymatic activation. The details of this critical activation process are now being actively studied, but are not clear at present. There is evidence that some caspases can process and activate other family members in vitro, but such interactions in dying cells are still not well resolved. The most convincing line of evidence that caspases are a common central element in diverse death pathways has come from the demonstration that a variety of inhibitors of these enzymes block many examples of cell death in vitro and in vivo (Henkart, 1996).

3. CASPASE INHIBITORS

As with other protease inhibitors, specific caspase inhibitors can be divided into two groups: protein inhibitors and small molecule inhibitors generally based on peptide substrates. There are currently two protein caspase inhibitors, both derived from viruses. Crm A is a cowpox-encoded protein which is a member of the serpin protease inhibitor family traditionally thought to be selective for serine proteases (Ray et al., 1992). Crm A is a good inhibitor of ICE (caspase-1) and appears to allow the virus to replicate while avoiding a severe inflammatory response. Crm A has an ICE-like recognition sequence in its reactive site loop, and as expected has considerably weaker ability to inhibit some other caspases such as caspase-3 (CPP-32), but could do so if expressed at high levels. Crm A seems particularly effective at blocking in vitro cell death via the Fas antigen, perhaps because Fas appears to utilize a particular route of activating downstream caspases via caspase-8 (FLICE) (Boldin et al., 1996; Muzio et al., 1996).

The other protein caspase inhibitor is the Baculovirus protein product p35, whose sequence is not homologous to others in the databases. p35 forms a tight 1:1 molar enzymatically inactive complex with caspases 1–4, but does not inhibit the asp-selective serine protease granzyme B (Bump et al., 1995). Inhibition depends on an internal DQMD sequence recognizable by most caspases, and although more is known about the mechanism of caspase inhibition by crmA than p35, the latter appears to be a more general caspase inhibitor than CrmA. Expression of p35 blocks developmental death in C.elegans and Drosophila as well as the experimental death of some cell lines (Hay et al., 1994; Sugimoto et al., 1994).

In order to test the ability of these protein caspase inhibitors to block cell death, they must be expressed in cells by transfection, which is feasible with cell lines but requires laborious construction of transgenic animals for in vivo systems. We have found that transient transfections are an excellent experimental approach in lymphoid cell lines, much preferable to construction of stable cell lines with consequent concerns of clonal variation. The principal limitation of transient transfections has always been that expression is typically achieved in fewer than 50% of the cells. However, cotransfection with markers such as β-galactosidase (Memon et al., 1995b) or surface anti-hapten antibody allows the relevant subpopulation to be examined exclusively, and this approach has allowed us to assess the effects of both crmA and p35 on a number of cell death systems.

Low molecular weight caspase inhibitors are another approach we have used, as described below. We feel strongly that this is often a simple and effective approach, although the choice of reagents and controls is critical. One of the first issues which must be considered is the toxicity of such inhibitors to the cells in question. Many reagents capable of inactivating proteases have additional non-specific chemical reactivity resulting in cell death. This is sometimes subtle, and we have seen cases where such inhibitors trigger a slow necrotic cell death, giving the potential impression that they block a more rapid apoptotic death when they are in fact just toxic. In our hands peptide chloromethyl ketones, which are excellent protease inhibitors, are generally toxic to T cells. The other problem with selective caspase inhibitors is that they generally have anionic amino acid side chains which retard their penetration into cells. This can be handled by electroporation or other means of permeabilization, or by synthesis of esters of these carboxylic acid side chains which are cleavable by cytoplasmic esterases. Available peptide fluoromethyl ketones with aspartic acid at the P1 position are synthesized as such esters, and we illustrate their general ability to block T cell death below. Fluoromethyl ketones are much less reactive than chloromethyl ketones, and in our experience they are remarkably non-toxic. It is critical to carry out control incubations with reagents bearing the same reactive group but with a non-asp amino acid at P1 if a claim is to be made for caspase inhibition, and we have used the cathepsin inhibitor ZFA-FMK as a control reagent for peptide-FMKs. The two caspase inhibitors in this class we have used are ZVAD-FMK (Armstrong et al., 1996) and Boc-D-FMK. While we do not know their reactivity with most caspases, we have tested their ability to inhibit ICE and CPP-32 (Sarin et al., 1996). ZVAD-FMK was designed to mimic the ICE cleavage site in pre-IL-1β, and is indeed an ICE inhibitor, but this reagent also inactivates CPP-32. Boc-D-FMK is also a CPP-32 inhibitor but not an ICE inhibitor.

4. MEASUREMENT OF CELL DEATH

. Before discussing the examples of the use of caspase inhibitors to block cell death, it is important to discuss the various cell death criteria which are currently being used and interpreted somewhat differently in different laboratories. In some cases parameters of nuclear damage considered as hallmark properties of apoptosis are exclusively monitored, with the interpretation that this is an adequate measure of death. While there is no question that DNA fragmentation or most other standard measures of apoptotic nuclear damage reflect an irreversible cell death process, such damage is clearly not required for some examples of apoptotic death, which have been shown to occur in enucleated cells. This is true for the Fas pathway as well as target cell death by the CTL granule exocytosis pathway (Nakajima et al., 1995). The experiments we will describe provide a clear example of why it is important to monitor a number of different readouts of cell death, testing the effect of caspase inhibitors on each independently. This allows one to assess whether or not caspases are upstream of the particular damage measured in the overall death pathway. In conjunction with other pharmacological reagents, this approach will allow the eventual construction of molecular pathways linking the various measures of lethal damage suffered by cells.

5. CASPASE INHIBITORS BLOCK MOST T CELL APOPTOTIC DEATH

5.1. ZVAD-FMK

We have measured the effect of peptide-FMK caspase inhibitors on many different T cell death systems. One of the most studied apoptosis systems is the immature T cells in the thymus, which are well known for the ability to undergo programmed cell death by numerous stimuli. As shown in Fig. 1, we have found that the compound ZVAD-FMK blocks spontaneous in vitro thymocyte death as well as that induced via four independent input pathways: Dexamethasone, which acts via the intracellular corticosteroid receptor; γ-irradiation, which induces DNA damage and sets off a p53-requiring death process; immobilized anti-CD3, which triggers a poorly understood but Fas-independent death process in these cells; and Fas, for which we now have a detailed input pathway model in which caspases play a prominent role (Boldin et al., 1996; Muzio et al., 1996). In this experiment we monitored both apoptotic nuclear morphology and cell viability by propidium iodide exclusion, and it can be seen that ZVAD-FMK blocks both readouts down to background. As a control reagent we used the cathepsin inhibitor ZFA-FMK, lacking an asp residue at P1, which had no significant effect. The striking aspect of this result is that the inhibition curves are so similar for all four input death pathways, implying that similar caspases are part of the death pathway in all these cases, plausibly the site of a common downstream element which is fed by distinct inputs. In any case this result had a big impact on our thinking about the role of caspases in T cell death, because we are unaware of any other pharmacological agent having a similar strong broad inhibitory activity, which is somewhat reminiscent of Bcl-2.

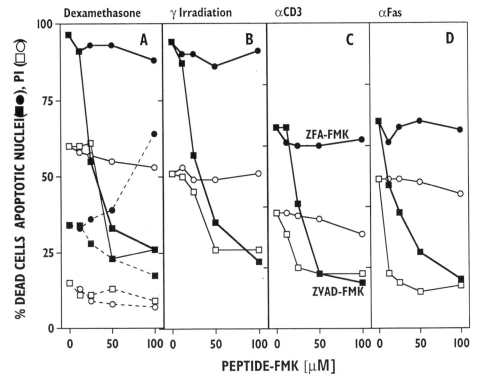

Figure 1. ZVAD-FMK blocks murine thymocyte death induced by different agents in vitro. The effects of ZVAD-FMK (squares) and ZFA-FMK (circles) on thymocyte death after 18 hours of culture in the presence of the following: A, spontaneous death (dotted lines) and dexamethasone (0.1μM, solid lines); B, γ–irradiation (5Gy); C, immobilized anti-CD3 (Tadakuma et al., 1990); D, immobilized anti-Fas (5μg/ml, Jo2 from PharMingen, Los Angeles, CA). Cell death was assessed by apoptotic nuclear morphology (closed symbols) and staining with propidium iodide (open symbols). A single experiment (representative of three) is shown.

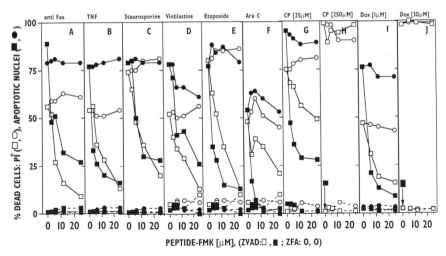

Figure 2. ZVAD-FMK blocks death of the human T cell tumor line Jurkat induced by various agents in vitro. 3×10^4 Jurkat cells were cultured for 18 hours in microtiter wells in the presence of the following: A, anti-Fas (100ng/ml, CH11 from Upstate Biotechnology, Lake Placid, NY); B, TNF (10ng/ml+ CHX 100ng/ml); C, Staurosporine (50nM) and the chemotherapeutic agents: D, vinblastine (5μg/ml); E, etoposide (5μg/ml); F, cytosine arabinoside (Ara C, 1μM); G, cisplatin (25 μM); H, cisplatin(250 μM); I, doxorubicin (1μM); and J, doxorubicin (30μM).. Cell death was assessed as described in Fig. 1. ZVAD-FMK (squares) or ZFA-FMK (circles) were added at the initiation of culture. This figure includes data from three experiments, with the background death in the absence of apoptotic stimuli shown in dotted lines.

T cell tumors provide another death system which can be triggered by numerous distinct input pathways. Some of these include the drugs currently used clinically for tumor chemotherapy, and it is clearly of interest to understand the molecular basis of such death. We have again considered the issue of caspase involvement using peptide-FMK inhibitors to block the in vitro apoptotic death of the human T cell tumor line Jurkat induced by various agents (Fig. 2). We have included four chemotherapeutic drugs with different molecular targets. At moderate concentrations all four agents induce apoptotic death after an overnight culture, and ZVAD-FMK specifically blocks both cytolysis and apoptotic nuclear damage. The inhibition curves are quite similar for the different drugs, although not as strikingly so as was the case for thymocytes. There is a clear tendency for cell death as monitored by propidium iodide to be less potently inhibited by ZVAD-FMK, but with the possible exception of 25μM cisplatin, inhibition is quite complete. Similar to earlier reports from other labs (Lennon et al., 1991), we find that high concentrations of some of these drugs induce a death which shows no apoptotic nuclear morphology. Interestingly, this death is not detectably blocked by ZVAD-FMK (panels H and J). Thus there is a correlation of death systems which show apoptotic nuclear damage and the ability to be inhibited by ZVAD-FMK.

While most of our efforts have gone into experiments with T lymphocyte-derived cells, we have been able to test ZVAD-FMK on some non-lymphoid cell death systems as well. Fig. 3 illustrates examples in which we examined the spontaneous death of mouse neutrophils in culture (panel A, recently been shown to be Fas-mediated (Liles et al., 1996)), staurosporine induced death of Chinese hamster ovary (CHO) cells (panel B) and ECM withdrawal induced death in human umbilical vein endothelial cells (panel C). The data indicate that these cell death systems are similar to lymphocytes in that they are also specifically blocked by ZVAD-FMK.

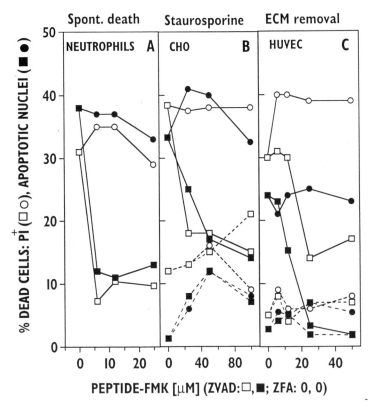

Figure 3. Inhibition of apoptotic death in various cell types by ZVAD-FMK. For all panels death was assayed after culture in the indicated concentrations of ZVAD (squares) or ZFA-FMK (circles). A. Mouse peritoneal neutrophils (Lagasse and Weissman, 1994) were elicited with thioglycollate, harvested after 4 hours, washed and allowed to adhere on Petri plates for 2 hours. Non-adherent cells were >95% neutrophils by morphology, and were cultured in complete medium. Death was assessed by nuclear morphology after 48 hours (solid symbols), and by propidium iodide staining after 72 hours (open symbols). B. CHO cells (2.5×10^5/ml in 24 well dishes, ATCC, Rockville, MD) were cultured in complete medium supplemented with .2mM proline. Chinese hamster ovary (CHO) cells were cultured for 18 hours in the presence (solid lines) or absence (dashed lines) of 100μM staurosporine. C. HUVEC (Clonetics, San Diego, CA) were cultured for 18 hours on wells precoated with gelatin (dashed lines) or albumin (solid lines) in defined serum-free medium as described (Meredith et al., 1993).

5.2. Boc-Asp-FMK

While the above examples and others from the literature (Jacobson et al., 1996; Cain et al., 1996; Pronk et al., 1996; Strasser et al., 1995; Fearnhead et al., 1995) illustrate the impressive ability of ZVAD-FMK to block apoptotic death processes, we have found other examples with T lymphocytes that make it clear that this reagent does not always block apoptotic death significantly. Interestingly, we have found that a related but truncated version of ZVAD-FMK, Boc-Asp-FMK (BD-FMK) generally is able to specifically inhibit death in these cases. Fig. 4 illustrates the example of resting peripheral T cells, which we had anticipated would behave similarly to thymocytes except that they are not as easy to kill. The data show that with all the agents which cause death, ZVAD-FMK gives a partial or negligible protection, measuring either apoptotic nuclei or death by trypan blue. Interestingly BD-FMK gives a rather potent and complete protection in all cases. In the case of steroid, ZVAD-FMK gives modest but significant protection from apoptotic nuclear damage, but its effect on death by trypan blue is negligible. One conclusion which we would draw from this data is that ZVAD-FMK probably does block death generally by virtue of its ability to inhibit ICE-like proteases with a preference for hydrophobic P4 amino acids, since, as described above: (1) BD-FMK fails to inhibit ICE; and (2) ZVAD-FMK also inhibits CPP-32-like caspases as does BD-FMK. These results point out that more work is needed to charactererize the specificity of reactivity of BD-FMK with the various caspases. Other examples where BD-FMK blocks the apoptotic death of T lymphoid cells better than of ZVAD-FMK include peripheral T cell blasts and most dramatically the T cell line CTLL-2 (Sarin et al., 1996).

All the above data argue for a correlation between caspase dependence and apoptotic phenotype, as might be predicted from a model in which apoptotic nuclear damage is a "useful epiphenomenon" linked to a common core death pathway including caspase activation and proteolysis. We found this model very appealing since it reconciled much of the data in the literature. Recently, however, we have realized that the simplest form of such a

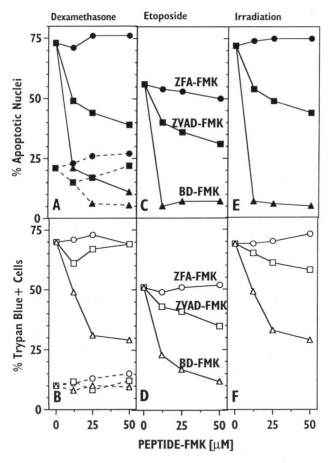

Figure 4. Inhibition of the apoptotic death of resting peripheral T cells by peptide-FMKs. T cells isolated from C57Bl/10 mice (the Jackson Laboratory, Bar Harbor, Maine) spleens were cultured without (dashed lines) or with dexamethasone (0.1 μM, A,B); etoposide (5 μM, C,D); and γ-irradiation (5 Gy prior to culture, E,F) in the indicated concentrations of ZVAD-FMK (squares), BD-FMK (triangles) or ZFA-FMK (circles) for 18 hours. Cell death was assessed by apoptotic nuclear morphology (A,C,E) and staining with trypan blue (Panels B,D,F).

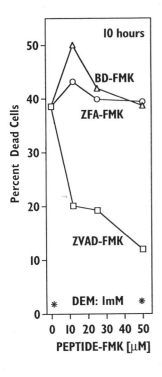

Figure 5. Inhibition of necrotic death by ZVAD-FMK in Jurkat cells. 3×10^4 Jurkat tumor cells were cultured for 10 hours with 1mM diethyl maleate (DEM, Sigma Chemical Co., St. Lois, MS) in the presence of ZVAD-FMK (squares), BD-FMK (triangles) or ZFA-FMK (circles). Cell death was assessed by a flow cytometric analysis of staining with propidium iodide. The percent apoptotic nuclei induced by DEM is indicated by an asterisk in the figure.

model is incorrect as it fails to predict cases where the correlation between caspase dependence and nuclear apoptotic phenotype breaks down. One such example is shown in Fig. 5, in which we examined the death of Jurkat cells treated with the membrane permeable thiol reagent diethyl maleate. This reagent is one of the limited number of ways we have found to induce an non-apoptotic death which is not very fast. On the basis of nuclear morphology and analysis of DNA content by FACS, we have found no nuclear apoptotic properties of this death at the reagent concentration used. This might then be considered a necrotic death. Nevertheless, ZVAD-FMK blocks this death quite nicely and specifically; curiously BD-FMK is inactive in this respect. Another example showing that caspase inhibitors can block non-apoptotic death comes from studies of agents inhibiting mitochondrial function in the pheochromocytoma cell line PC12 (Shimizu et al., 1996). Thus it may be that some death pathways involving caspases fail to trigger apoptotic nuclear damage.

6. LYMPHOCYTE-MEDIATED CYTOTOXICITY

6.1. Effect of Peptide-FMK Caspase Inhibitors

Another, more interesting, example of a lack of correlation between caspase sensitivity and apoptotic phenotype comes from our studies of lymphocyte-mediated cytotoxicity. For some years now our lab has pursued the question of how target cells die when cytotoxic lymphocytes release their granule contents onto the surface of the bound target cell. We have proposed the model that granzymes are the actual mediators of target death and are trying to define the complete molecular pathway of this death. After showing that target nuclei are not required (Nakajima et al., 1995), we have considered the hypothesis that granzymes activate caspases which in turn leads to target cell death. Fig. 6 shows the

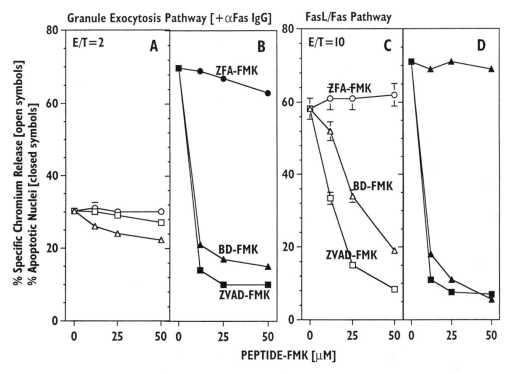

Figure 6. Effect of peptide-FMKs on CTL killing of Jurkat cells. A, B: TNP modified Jurkat tumor cells were labeled with the lipophilic membrane dye Di-I-C_{16} (Molecular Probes, Eugene, OR) and incubated for 4 hours with allogeneic murine CTL redirected with 100 ng/ml αCD3xαTNP heteroconjugate (Segal and Wunderlich, 1988). Soluble anti Fas IgG (10μg/ml, DX2, PharMingen) was added at the initiation of culture to block the Fas pathway in these experiments. C, D: Di-I-C_{16} labeled Jurkat target cells were incubated with Fas ligand bearing d11S CTL (Rouvier et al., 1993) effector cells for 4 hours. All assays were set up in 96 well microtiter plates and ZVAD-FMK (squares), BD-FMK (triangles) or ZFA-FMK (circles) were added at the initiation of culture. Cell death was assayed by the chromium release assay of cytotoxicity (A and C) and apoptotic nuclear morphology (B and D) of Jurkat target cells was assessed as described in Fig. 1. Each data point in panels A and C represents the mean±SD of counts from three individual wells.

interesting result obtained when we used peptide-FMKs to inhibit CTL killing of Jurkat target cells in standard four hour assays. These cells bear functional Fas and thus can be killed by both of the two pathways used by CTL for short-term in vitro cytotoxicity (Henkart, 1994). Given the result in Fig. 3 that ZVAD-FMK blocks Jurkat death induced by anti-Fas antibody, it is no surprise that it and BD-FMK block both apoptotic nuclear morphology and lysis when Fas is crosslinked by Fas Ligand on the surface of a CTL (6 C and D). However the interesting result is obtained when the Fas pathway is shut down with non-crosslinking anti-Fas so that all the killing goes by granule exocytosis. In that case ZVAD-FMK and BD-FMK block apoptotic nuclear morphology induced in Jurkat much like the result for the Fas pathway. However, lysis of Jurkat as measured by ^{51}Cr release is not significantly blocked by ZVAD-FMK or BD-FMK. The data shown are from an experiment in which we tried to maximize the chances of seeing such inhibition by using a low E/T, reasoning that the lower levels of lysis should be more easily inhibited. We have used this approach with other target cells including thymocytes and gotten the same results: target cell lysis is unaffected by the peptide-FMK caspase inhibitors while they do block nuclear damage assessed by morphology and DNA release.

6.2. Effect of Baculovirus p35

Because it is possible that some caspases may not be blocked by peptide-FMK reagents we thought it was important to examine the effect of Baculovirus p35, because of its very general ability to block caspases. When p35 was transiently expressed in Jurkat, lysis induced by anti-Fas antibody, as monitored by the release of cotransfected β-galactosidase, was dramatically blocked (Fig. 7B). When CTL were used, p35 expression completely blocked that portion of target lysis which was also blocked by IgG anti-Fas antibody (Fig. 7A). However, the remaining major Fas-independent portion of target lysis which is due to granule exocytosis is unaffected by p35. We have used other target cells which express no Fas and confirmed this result, and also shown that target apoptotic damage induced by the granule exocytosis pathway is blocked by p35. These data thus confirm the results obtained with peptide-FMKs. They are compatible with recent findings that granzyme B (as well as Fas crosslinking) can activate caspases, and we think it likely that this is responsible for CTL-induced target nuclear damage. However, one cannot deny that a target cell which is lysed is dead even if it shows no apoptotic nuclear damage, and we believe that our data strongly imply that the granule exocytosis death pathway has nothing to do with caspases.

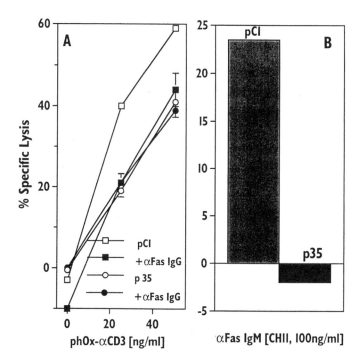

Figure 7. Baculovirus protein p35 inhibits Fas mediated but not CTL killing of Jurkat cells. Jurkat tumor cells were transiently transfected with 5μg cmv-β–galactosidase, 10μg pCI-pHook and 10μg pCI-p35 (squares) or 10μg pCI empty vector plasmid (circles) as described (Memon et al., 1995a). Following an overnight culture in complete medium, live cells were harvested by centrifugation over Ficoll. Transiently transfected cells were either, A: assayed for susceptibility to lysis by murine allogeneic CTL, redirected for killing with the indicated concentration of phOx-αCD3 antibody in the presence (closed symbols) or absence (open symbols) of a soluble αFas IgG blocking antibody (10μg/ml DX2) or B: cultured in the presence of a cytotoxic αFas IgM antibody (100 ng/ml, CH11). After 6 hours the cells were harvested and assayed for residual β-galactosidase activity as described (Memon et al., 1995a). In panel A, each data point represents a mean±SD of counts from three separate wells.

Table 1. Alternative systems for classifying cell death

	Nuclear phenotype			
	Necrotic		Apoptotic	
	Cell	Stimulus	Cell	Stimulus
Caspase dependent	Jurkat	Diethyl maleate	Jurkat	Fas, TNF, staurosporine, cisplatin (low), etoposide, doxorubicin (low), ara C
			Thymocytes	Fas, steroid, αCD3, DNA damage
			Peripheral T	Steroid, DNA damage
Caspase independent	Jurkat	Cisplatin (high) Doxorubicin (high)	Jurkat	CTL granule exocytosis

7. CONCLUSIONS

We began our studies with caspase inhibitors thinking that they would confirm the appealing model that the apoptotic characteristics of cell death reflect the operation of a common core molecular death pathway into which various input pathways funnel (Henkart, 1996). This would be reflected in a correlation between the apoptotic character of cell death and its ability to be blocked by caspase inhibitors, and Table 1 shows that indeed we have found considerable evidence which supports such a correlation. For us, however, the most interesting lessons from our studies have come from the exceptions to this correlation, because they point out that if one is trying to understand molecular death pathways, the current conceptual framework of apoptosis may be fraught with serious problems. Caspase inhibitors are newly available tools which allow a molecular approach to defining death pathways, which clearly need to be worked out in further detail. When we do so it may turn out that the caspase-dependent necrotic deaths actually use a different death pathway than the caspase-dependent apoptotic deaths, although this seems unlikely at this point. The case of cytotoxic lymphocytes provides a clear lesson that cell death is not necessarily simple, and it is possible to be misled if a number of different death read-outs are not measured. Other molecular components of the common caspase-dependent death pathway will be identified within the next few years, and specific inhibitors of their function will provide further tools for defining more complete molecular pathways of cell death in various physiologically interesting and clinically important systems. While major advances have been made, there are still important gaps in our knowledge of death pathways which will provide us with future challenges.

REFERENCES

Alnemri, E.S., Livingston, D.J., Thornberry, N.A., Wong, W.W., and Yuan, J. (1996). Human ICE/CED-3 protease nomenclature. Cell 87, 171

Armstrong, R.C., Aka, T., Xiang, J., Gaur, S., Krebs, J.F., Hoang, K., Bai, X., Korsmeyer, S.J., Karanewsky, D.S., Fritz, L.C., and Tomaselli, K.J. (1996). Fas-induced activation of the cell death-related protease CPP-32 is inhibited by Bcl-2 and by ICE family protease inhibitors. J. Biol. Chem. 271, 16850–16855.

Boldin, M.P., Goncharov, T.M., Goltsev, Y.V., and Wallach, D. (1996). Involvement of MACH, a novel MORT1/FADD-interacting protease, in Fas/APO-1- and TNF receptor-induced cell death. Cell 85, 803–815.

Bump, N.J., Hackett, M., Hugunin, M., Seshagiri, S., Brady, K., Chen, P., Ferenz, C., Franklin, S., Ghayur, T., Li, P., et al. (1995). Inhibition of ICE family proteases by baculovirus antiapoptotic protein p35. Science 269, 1885–1888.

Cain, K., Inayat-Hussain, S.H., Couet, C., and Cohen, G.M. (1996). A cleavage-site-directed inhibitor of interleukin 1β-converting enzyme-like proteases inhibits apoptosis in primary cultures of rat hepatocytes. Biochem. J. 314, 27–32.

Fearnhead, H.O., Dinsdale, D., and Cohen, G.M. (1995). An interleukin-1-beta converting enzyme-like protease is a common mediator of apoptosis in thymocytes. FEBS Lett. *375*, 283–288.

Hay, B.A., Wolff, T., and Rubin, G.M. (1994). Expression of baculovirus p35 prevents cell death in Drosophila. Development *120*, 2121–2129.

Hengartner, M.O. and Horvitz, H.R. (1994). Programmed cell death in Caenorhabditis elegans. Curr. Opin. Genet. Dev. *4*, 581–586.

Henkart, P.A. (1994). Lymphocyte-mediated cytotoxicity: Two pathways and multiple effector molecules. Immunity *1*, 343–346.

Henkart, P.A. (1996). ICE Family Proteases: Mediators of all apoptotic cell death? (Review). Immunity *4*, 195–201.

Jacobson, M.D., Weil, M., and Raff, M.C. (1996). Role of Ced3/ICE-family proteases in staurosporine-induced programmed cell death. J. Cell Biol. *133*, 1041–1051.

Lagasse, E. and Weissman, I.L. (1994). bcl-2 inhibits apoptosis of neutrophils but not their engulfment by macrophages. J. Exp. Med. *179*, 1047–1052.

Lennon, S.V., Martin, S.J., and Cotter, T.G. (1991). Dose-dependent induction of apoptosis in human tumor cell lines by widely diverging stimuli. Cell Prolif. *24*, 203–214.

Liles, W.C., Kiener, P.A., Ledbetter, J.A., Aruffo, A., and Klebanoff, S.J. (1996). Differential expression of Fas (CD95) and Fas Ligand on normal human phagocytes: Implications for the regulation of apoptosis in neutrophils. J. Exp. Med. *184*, 429–440.

Memon, S.A., Moreno, M.B., Petrak, D., and Zacharchuk, C.M. (1995a). Bcl-2 blocks glucocorticoid- but not Fas- or activation-induced apoptosis in a T cell hybridoma. J. Immunol. *155*, 4644–4652.

Memon, S.A., Petrak, D., Moreno, M.B., and Zacharchuk, C.M. (1995b). A simple assay for examining the effect of transiently expressed genes on programmed cell death. J. Immunol. Methods *180*, 15–24.

Meredith, J.E., Fazeli, B., and Schwartz, M.A. (1993). The extracellular matrix as a cell survival factor. Mol. Biol. Cell *4*, 953–961.

Muzio, M., Chinnaiyan, A.M., Kischkel, F.C., O'Rourke, K., Shevchenko, A., Ni, J., Scaffidi, C., Bretz, J.D., Zhang, M., Gentz, R., Mann, M., Krammer, P.H., Peter, M.E., and Dixit, V.M. (1996). FLICE, a novel FADD-homologous ICE/CED-3-like protease, is recruited to the CD95 (Fas/Apo-1) death-inducing signaling complex. Cell *85*, 817–827.

Nakajima, H., Golstein, P., and Henkart, P.A. (1995). The target cell nucleus is not required for cell-mediated granzyme- or Fas-based cytotoxicity. J. Exp. Med. *181*, 1905–1909.

Pronk, G.J., Ramer, K., Amiri, P., and Williams, L.T. (1996). Requirement of an ICE-like protease for induction of apoptosis and ceramide generation by REAPER. Science *271*, 808–810.

Ray, C.A., Black, R.A., Kronheim, S.R., Greenstreet, T.A., Sleath, P.R., Salvesen, G.S., and Pickup, D.J. (1992). Viral inhibition of inflammation: Cowpox virus encodes an inhibitor of the interleukin-1B converting enzyme. Cell *69*, 597–604.

Rouvier, E., Luciani, M.F., and Golstein, P. (1993). Fas involvement in Ca+2-independent T cell-mediated cytotoxicity. J. Exp. Med. *177*, 195–200.

Sarin, A., Wu, M.-L., and Henkart, P.A. (1996). Different ICE-family protease requirements for the apoptotic death of T lymphocytes triggered by diverse stimuli. J. Exp. Med. *184*,

Segal, D.M. and Wunderlich, J.R. (1988). Targeting of Cytotoxic Cells with Heterocrosslinked Antibodies. Cancer Invest. *6*, 83–92.

Shimizu, S., Eguchi, Y., Kamiike, W., Waguri, S., Uchiyama, Y., Matsuda, H., and Tsujimoto, Y. (1996). Bcl-2 blocks loss of mitochondrial membrane potential while ICE inhibitors act at a different step during inhibition of death induced by respiratory chain inhibitors. Oncogene *13*, 21–29.

Strasser, A., Harris, A.W., Huang, D.C.S., Krammer, P.H., and Cory, S. (1995). Bcl-2 and Fas/APO-1 regulate distinct pathways to lymphocyte apoptosis. EMBO J. *14*, 6136–6147.

Sugimoto, A., Friesen, P.D., and Rothman, J.H. (1994). Baculovirus *p35* prevents developmentally programmed cell death and rescues a *ced-9* mutant in the nematode *Caenorhabditis elegans*. EMBO J. *13*, 2023–2028.

Tadakuma, T., Kizaki, H., Odaka, C., Kubota, R., Ishimura, Y., Yagita, H., and Okumura, K. (1990). CD4+CD8+ thymocytes are susceptible to DNA fragmentation induced by phorbol ester, calcium ionophore and anti-CD3 antibody. Eur. J. Immunol. *20*, 779–784.

7

CELL VOLUME REGULATION, IONS, AND APOPTOSIS

Carl D. Bortner, Francis M. Hughes, Jr., and John A. Cidlowski[*]

The Laboratory of Signal Transduction
National Institute of Environmental Health Sciences, NIH
Research Triangle Park, North Carolina 27709

1. INTRODUCTION

It is generally accepted that all cells have the ability to undergo an internally controlled cell suicide process known as apoptosis, or programmed cell death, in response to a given stimulus or environmental agent[39]. The apoptotic process efficiently removes or eliminates a population of unwanted cells from the body at a given time in the absence of an inflammatory response. This mode of cell death has been observed during development[25,37,40], in the immune system[2], and in the progression of both AIDS[21,33,34] and cancer[24,35,38]. Apoptosis is characterized by a distinct set of morphological and biochemical features including cell shrinkage, nuclear condensation, proteolysis, internucleosomal DNA cleavage, and apoptotic body formation[27]. Over the past 20 years, much attention has focused on the biochemical aspects of apoptosis, including the intracellular signals leading to cell death, the enzymes involved in both protein and DNA degradation, and the role several apoptotic modulating genes play to control or inhibit cell death[12]. However, of all the characteristics which define this mode of cell death, the observation that the cells shrink or lose volume during apoptosis has remained relatively unexplored.

This unique and distinctive feature of cell shrinkage during apoptosis has been observed from the very first reports of apoptosis[19,44]. Subsequently, the loss of cell volume during programmed cell death has been observed in all well-defined cases of apoptosis, thus likely reflecting a key component of the cell death process. The loss of cell volume or cell shrinkage during apoptosis is exclusively observed during this mode of cell death. When cells die by an accidental cell death process known as necrosis, the cells swell. This cellular swelling observed during necrosis is caused by the early loss of energy and cell membrane integrity from the cell permitting the movement of both ions, mostly sodium, and water into the cell. Necrotic cells eventually lyse, in turn triggering an inflammatory response, which is not observed during apoptosis. In contrast, during apoptosis, energy levels remain high in the cell and the cells retain their membrane integrity until very late

---

[*] Corresponding author.

in the cell death process. The cells shrink and are eventually phagocytized by neighboring cells to efficiently eliminate the unwanted cells from the body. Interestingly, the mechanism behind this loss of cell volume from apoptotic cells is currently unknown.

2. CELL VOLUME AND REGULATION OF CELL FUNCTION

Recently, the idea that changes in cell volume can signal and effect intracellular mechanisms such as cellular metabolism and gene expression has come to light[13]. In hepatocytes for example, the most extensively studied model system for volume effects on cellular function, hypoosmotic cell swelling acts as an anabolic signal, whereas hyperosmotic cell shrinkage is catabolic. Additionally, it has been shown that hormones, oxidative stress, amino acids, and other nutrients can also act on the cellular hydration state to alter cell function[14]. Hormones are known to modify the activities of ion transporters and channels in hepatocytes which is reflected in changes in the cell membrane and membrane potential[30]. However, the precise role that hormone-induced cell volume changes play has yet to be established. Several examples of modified cellular function in response to the cellular hydration state exists in other cell types, including studies done on lymphocytes and macrophages[43]. Decreasing the medium osmolarity was shown to increased the rates of glutamine metabolism and glycolysis in both rat lymphocytes and macrophages. Conversely, increasing the medium osmolarity had the opposite effect. Increasing amounts of evidence suggests that cell volume may act as a signalling system for metabolic control. Therefore, cell volume probably plays a more significant role in cellular physiology then just maintaining cell size.

3. CONTROL OF CELL VOLUME AND VOLUME REGULATION DURING APOPTOSIS

Most mammalian cells are capable of controlling their own cell volume in response to various anisotonic conditions[1,16]. Although the mechanisms which allows for volume recovery differ among various cell types, they all involve either the transport of ions or the use of diverse organic osmolytes. In general, when cells are exposed to hypotonic conditions, they immediately swell or increase in cell size. However, after a period of time, many cells can compensate for this increase in cell volume by the activation of specific ion transport mechanisms collectively known as a regulatory volume decrease (RVD) response. During this response, the cells lose ions, particularly K^+ and Cl^-, causing the obligatory movement of water out of the cells, which allows these cells to eventually achieve a near normal cell volume. In contrast, when cells are exposed to hypertonic conditions, they immediately shrink or decrease in cell size. Again, after a period of time, most cells can compensate for this decrease in cell volume by the activation of specific ion transport mechanisms collectively known as a regulatory volume increase (RVI) response. During this response, the cells gain ions, particularly Na^+ and Cl^- (with the Na^+ eventually being exchanged for K^+ by the action of the Na^+/K^+ ATPase pump), which allows for the movement of water into the cells and the recovery of a near normal cell volume. Although the ionic transporters and channels associated with these responses are known for a few cell types, how these responses are activated is not well understood.

Lymphocytes, specifically T lymphocytes, have become one of the most well characterized model systems for studying apoptosis. Interestingly, lymphocytes have also been one the most well studied model systems for investigating cell volume regulation. The RVD response in T lymphocytes occurs by the loss of both K^+ and Cl^- through individual channels[10]. Therefore, this RVD response to hypotonic conditions returns these cells from a swollen state to a near normal cell volume. Interestingly, T lymphocytes when exposed to hypertonic conditions shrink and remain in a state of reduced cell volume [15,36]. Figure 1A illustrates the response of thymic lymphocytes to changes in their cell volume. Surprisingly, these T lymphocytes do contain the mechanisms for an active RVI response, however this is only observed when the cells are first primed by a prior treatment under hypotonic conditions [11]. This secondary RVI response occurs by the activation of a Na^+/H^+ exchanger coupled to a Cl^-/HCO_3^- exchanger. In contrast, most mammalian cells can respond to changes in cell volume with either an RVI or RVD response (Figure 1B).

As described above, thymocytes do not undergo an RVI response when cultured under hypertonic conditions. We have recently examined if this loss of cell volume which occurs under hypertonic conditions can trigger apoptosis in an immature T-cell line designated S49 Neo [8]. When S49 Neo cells were exposed to either hypertonic mannitol, sucrose, or NaCl added to normal culture media, a rapid loss of viability was observed in both a time and concentration dependent manner. Total DNA isolated from these cells showed the presence of internucleosomal DNA cleavage, characteristic of apoptosis. Using flow cytometry, we also showed a loss of cell volume as a population of cells was observed with a reduced forward light scatter (a direct measure of cell size) and the presence of a subdiploid

A   Regulation of Cell Volume in Thymic Lymphocytes

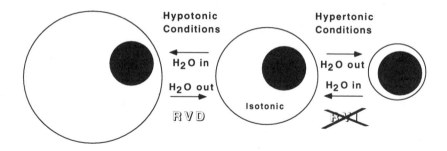

B   Regulation of Cell Volume in Most Cells

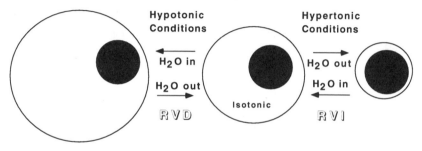

Figure 1. Regulation of cell volume in mammalian cells. Thymic lymphocytes (A) can activate an RVD response when exposed to hypotonic conditions; however, they cannot activate an RVI response under hypertonic conditions. In contrast, (B) most mammalian cells can activate either an RVD or an RVI response to anisotonic conditions.

Table 1. Response of lymphoid cells to mannitol-induced hypertonicity

	S49 Neo cells	CEM-C7 cells	Primary thymocytes
Initial decrease in cell volume	Yes	Yes	Yes
Compensatory volume regulation	No	No	No
DNA degradation	Yes	Yes	Yes
Apoptosis	Yes	Yes	Yes

peak of DNA in cells stained with propidium iodide, again characteristic of apoptosis when used in conjunction with other apoptotic assessment techniques. Interestingly, when these S49 Neo cells were exposed to hypotonic conditions where the cells swell, no cell death was observed, suggesting that a specific loss of cell volume is required to induce apoptosis and not just a general change in cell volume. We extended these studies to include other thymic cells which do not activate an RVI response and observed a similar apoptotic response (Table 1). Interestingly, when non-lymphoid cells, such as COS, HeLa, GH$_3$, and L-cells, were cultured under identical hypertonic conditions as the lymphoid cells, no apoptotic response was observed (Table 2). Using flow cytometry, we observed that these cells showed an RVI response to the hypertonic treatment. This suggests that the presence of an RVI response may act as the first line of defense to protect cells against apoptosis. Therefore, the RVI response in these cells, which are known to undergo apoptosis in response to other agents must be either inhibited or overridden for the programmed cell death process to ensue. Hypertonically induced apoptosis is not limited to lymphoid cells. Hepatocytes (Bortner and Cidlowski, unpublished results) and human neuroblastoma cells [28] were also shown to undergo programmed cell death in response to hypertonic conditions.

4. THE ROLE OF CELL SHRINKAGE DURING APOPTOSIS

When apoptotic cells are examined by flow cytometry on a forward scatter versus side scatter dot plot to compare the relationship between cell size and cell density respectively, two distinct populations are observed [8]. This observation that apoptotic cells appear as a separate population of cells based on cell size has been universally observed [41,45] and the loss of cell volume during apoptosis has been shown to occur in distinct stages [20,31,32]. Additionally, we have shown that only cells which lose cell volume have degraded their DNA [8]. This correlation between the shrunken population and degraded DNA has been shown to occur in other cells treated with a variety of apoptotic agents (Bortner and Cidlowski, unpublished results). Therefore, the loss of cell volume during apoptosis plays a very critical role during the programmed cell death process. However, the question still remains as to how this loss of cell volume occurs.

The intracellular environment of most mammalian cells is high in potassium and low in sodium content, the complete opposite of what is observed in the extracellular environment. Water can flow freely through the cell membrane, however the amount of water that can enter a cell is tightly controlled by the cells ionic composition which occurs through

Table 2. Response of non-lymphoid cells to mannitol-induced hypertonicity

	L-cells	COS cells	HeLa cells	GH3 cells
Initial decrease in cell volume	Yes	Yes	Yes	Yes
Compensatory volume regulation	Yes	Yes	Yes	Yes
DNA degradation	No	No	No	No
Apoptosis	No	No	No	No

the activation of various transporters and exchangers. Therefore, the loss of cell volume during apoptosis may be in response to a loss of ions from the cell. Recently, several investigators have begun to investigate the role ions play during apoptosis. Since cells maintain a high intracellular potassium concentration, the observed loss of cell volume during apoptosis may occur by the loss of this ion from the cell. Cells are bathed in a milieu of low potassium, therefore a natural gradient exists for the flow of potassium and water out of the cell. When human eosinophils are cultured in the absence of any cytokines they undergo apoptosis, however the addition of a potent K^+ channel blocker 4-aminopyridine inhibits the loss of cell volume by approximately 60%[5]. Similar results were observed in the presence of two other K^+ channel blockers, sparteine and quinidine, but unfortunately the biochemical characteristics of apoptosis were not assessed in the presence of these channel inhibitors. Benson et al.[6] also reported a loss of intracellular potassium during dexamethasone-induced apoptosis of CEM-C7A cells, however calculations of intracellular water suggest no change in potassium concentration during this efflux of ions from the cells as the loss of cell volume balanced the loss of potassium ions. In contrast, L cells treated with the DNA topoisomerase inhibitor VP-16 showed a strong decrease in potassium in a fraction of the cells which had loss cell volume and retained their membrane integrity[3]. Therefore the loss of potassium from cells undergoing apoptosis may allow for the characteristic cell shrinkage associated with apoptosis, but may also play a more fundamental role during the cell death process.

5. CONTROL OF ENZYMATIC ACTIVITY BY POTASSIUM

If K^+ is selectively lost from an apoptotic or preapoptotic cell, it would significantly change the intracellular environment in which the cell death machinery operates. Studies addressing the effects of the intracellular ionic milieu on the activity of apoptotic pathways are in their infancy, but early results suggest that their influence may be considerable. For example, we[17] and others[18] have explored the effects of intracellular levels of K^+ on the degradation of genomic DNA onto oligonucleosomal fragments, an enzymatic event that is widely used as a marker of this process. Figure 2 shows that 150 mM KCl (equivilent to normal intracellular levels) can completely inhibit the oligonucleosomal fragments pro-

KCl (mM)

0   150

Figure 2. Autodigestion of thymocyte DNA is inhibited by 150 mM KCl. Thymocyte nuclei were prepared from adrenalectomized rats and incubated in 50 mM Tris containing 1 mM $MgCl_2$, 1 mM $CaCl_2$ in the presence or absence of 150 mM KCl. The reaction was stopped after 1.5 hours by the addition of EDTA and treatment with proteinase K. DNA was then extracted, precipitated and electrophoresed on a 1.8% agarose gel followed by staining with ethidium bromide.

duced in an in vitro model. In addition, Jonas et al.[18] recently found that permeablizing cells with a monovalent ion pore produced by alpha-toxin induces DNA degradation and this degradation could be inhibited by culture in a medium containing high K^+ levels. Thus it is likely that intracellular levels of K^+ exert tonic inhibitory effects on the apoptotic nuclease(s), preventing inappropriate activation and chromatin destruction in a normal cell.

Much attention has also focused recently on the role of proteases in apoptosis, particularly those enzymes related to interleukin-1β converting enzyme (ICE)[4,23,26]. ICE was initially isolated for its ability to cleave proIL-1 to IL-1 prior to secretion of this cytokine[7,9,22]. Subsequently, it was found that ICE has significant homology to ced3, a C. elegans gene known to play an intimate role in the cellular death of this species[29]. The importance of these enzymes to apoptosis was confirmed when it was shown that overexpression of ICE or ICE related proteases induces cell death while inhibition of these enzymes suppresses apoptosis in a variety of cell lines and in response to a variety of agents (see [23] and references therein). Thus it appears that this class of proteases plays a central role in the apoptotic process. The activity of ICE can be conveniently measured as the amount of proIL-1 versus mature IL-1 in the intracellular and extracellular fluid. In a recent study Walev et al.[42] demonstrated that intracellular K^+ depletion (through an alpha-toxin generated monovalent ion pore) stimulated the conversion of proIL-1 to IL-1. Interestingly, when these cells were placed in medium containing high levels of K^+, this conversion was inhibited, suggesting that K^+ levels can exert an inhibitory influence on ICE activity. The effects of K^+ on other members of the ICE family remain to be explored but the results with ICE suggest that K^+ may have inhibitory effects on these enzymes as well.

6. ROLE OF K^+ DURING APOPTOSIS

As discussed above, studies from a variety of different areas are beginning to shed light on the importance of K^+ to cell death and the results suggest that this ion plays a multifaceted role in the apoptotic process. In a normal nondying cell high intracellular K^+ maintains appropriate cell volume while simultaneously acting on several key apoptotic enzymes to suppress activity. This tonic inhibition provides built-in check points to insure against unwanted distal consequences (such as degradation of the genome) if the proximal portions of the pathway are inadvertently or inappropriately activated. We propose that when a signal triggers the apoptotic program one of the initial changes in a common death pathway, regardless of the nature of the signal, is the selective loss of intracellular K^+. As K^+ is removed it draws water out with it, forcing the cell to shrink. The loss of intracellular volume, coupled with the loss of K^+ ions, initially results in no net change in K^+ concentration from that originally present in the nondying cell. However, as the process of K^+ extrusion and cell shrinkage continues, the osmotic effects of the impermeant molecules in the cell (proteins, sugars, etc.) soon become significant and less water follows each K^+ ion. This change in the number of water molecules leaving the cell per K^+ ion leads to a real decrease in the intracellular concentration of K^+ in the shrunken cell. The tonic inhibition of enzyme activity normally exerted by high K^+ is consequently relieved, allowing the next phase of the apoptotic cascade (protease activation, nuclease activation, etc.) to proceed and bring about the final demise of the cell.

In conclusion, we present a general model of the apoptotic process in figure 3. In this model, volume and ionic changes are early events and may represent one of the first steps on a common apoptotic program in all cells. These changes alter the intracellular environment so that it becomes permissive to the enzymatic activities which follow. Future

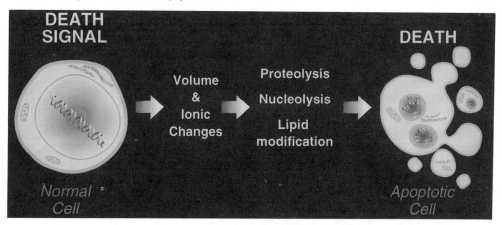

Figure 3. Model depicting the proposed sequence of events following exposure of a cell to a death signal. We propose that the first common change in all apoptotic cells is a loss of K^+ concomitant with a loss of cell volume (i.e. cell shrinkage). The lower intracellular K^+ then provides a permissive environment for subsequent enzyme activities such as proteases and nucleases which eventually lead to recognition and engulfment of the apoptotic cell by phagocytes.

research will hinge upon the identification of the mechanisms controlling the K^+ efflux in a dying cell and if these mechanisms are consistent across other apoptotic models. Such discoveries may become the targets of future pharmacological intervention.

REFERENCES

1. Al-Habori, M. Cell volume and ion transport regulation. Int. J. Biochem. 1994; 26:319–34.
2. Allen PD, Bustin SA, and Newland AC. The role of apoptosis (programmed cell death) in haemopoiesis and the immune system. Blood Rev. 1993; 7:63–73.
3. Barbiero G, Duranti F, Bonelli G, Amenta JS, and Baccino FM. Intracellular ionic variations in the apoptotic death of L cells by inhibitors of cell cycle progression. Expt. Cell Res. 1995; 217:410–8.
4. Barinaga M. Cell suicide: by ICE, not fire. Science. 1994; 263:754–6.
5. Beauvais F, Michel L, and Dubertret L. Human eosinophils in culture undergo a striking and rapid shrinkage during apoptosis. Role of K^+ channels. J. Leukoc. Biol. 1995; 57:851–5.
6. Benson RSP, Heer S, Dive C, and Watson AJM. Characteristics of cell volume loss in CEM-C7A cells during dexamethasone-induced apoptosis. Am. J. Physiol. 1996; 270:C1190–1203.
7. Black RA, Kronheim SR and Sleath PR. Activation of interleukin-1 beta by a co-induced protease. FEBS Lett. 1989; 247:386–390.
8. Bortner CD, and Cidlowski JA. The absence of volume regulatory mechanisms contributes to the rapid activation of apoptosis in thymocytes. Am. J. Physiol. 1996; 271:C950–61.
9. Cerretti DP, Kozlosky CJ, Mosley B, Nelson N, Van-Ness K, Greenstreet TA, March CJ, Kronheim SR, Druk T, Cannizzaro LA, Huebner K, and Black RA. Molecular cloning of the interleukin-1 beta converting enzyme. Science. 1992; 256:97–104.
10. Cheung RK, Grinstein S, and Gelfand EW. Volume regulation by human lymphocytes; Identification of differences between the two major lymphocyte subpopulations. J. Clin. Invest. 1992; 70:632–38.
11. Grinstein S, Clarke CA, and Rothstein A. Activation of Na^+/H^+ exchange in lymphocytes by osmotically-induced volume changes and by cytoplasmic acidification. J. Gen. Physiol. 1983; 82:619–38.
12. Hale AJ, Smith CA, Sutherland LC, Stoneman VEA, Longthorne VL, Culhane AC, and Williams GT. Apoptosis: molecular regulation of cell death. Eur. J. Biochem. 1996; 236:1–26.
13. Haussinger D. The role of cellular hydration in the regulation of cell function. Biochem. J. 1996; 313:697–710.
14. Haussinger D, Lang F, and Gerok W. Regulation of cell function by the cellular hydration state. Am. J. Physiol. 1994; 267:E343–55.
15. Hempling HG, Thompson S, and Dupre A. Osmotic properties of human lymphocytes. J. Cell. Physiol. 1977; 93:293–302.

16. Hoffmann EK. Volume regulation in cultured cells. Current Topics in Membranes and Transport 1987; 30:125–80.

17. Hughes FM Jr. and Cidlowski JA. Submitted. 1997.

18. Jonas D, Walev I, Berger T, Liebetrau M, Palmer M and Bhakdi S. Novel path to apoptosis: small transmembrane pores created by staphylococcal alpha-toxin in T lymphocytes evoke internucleosomal DNA degradation. Infect Immun. 1994; 62:1304–12.

19. Kerr JFR., Wyllie AH, and Currie AR. Apoptosis: a basic biological phenomenon with wide-ranging implications in tissue kinetics. Br. J. Cancer 1972; 26:239–57.

20. Klassen NV, Walker PR, Ross CK, Cygler J, and Lach B. Two-stage cell shrinkage and the OER for radiation-induced apoptosis of rat thymocytes. Int. J. Radiat. Biol. 1993; 64:571–81.

21. Kornbluth RS. Significance of T cell apoptosis for macrophages in HIV infection. J. Leukoc. Biol. 1994; 56:247–56.

22. Kostura MJ, Tocci MJ, Limjuco G, Chin J, Cameron P, Hillman AG, Chartrain NN and Schmidt JA. Identification of a monocyte specific pre-interleukin 1 beta convertase activity. Proc Natl Acad Sci USA. 1989; 85:5227–31.

23. Kumar S. ICE-like proteases in apoptosis. Trends Biochem Sci. 1995; 20:198–202.

24. Lee JM, and Bernstein A. Apoptosis, cancer, and the p53 tumour suppressor gene. Cancer Metastasis Rev. 1995; 14:149–61.

25. Lo AC, Houenou LJ, and Oppenheim RW. Apoptosis in the nervous system: morphological features, methods, pathology, and prevention. Arch. Histol. Cytol. 1995; 58:139–49.

26. Martin SJ and Green DR. Protease activation during apoptosis, death by a thousand cuts? Cell. 1995; 82:349–52.

27. Martin SJ, Green DR, and Cotter TG. Dicing with death: dissecting the components of the apoptotic machinery. Trends Biochem. Sci. 1994; 19:26–30.

28. Matthews CC, and Feldman EL. Insulin-like growth factor 1 rescues SH-SY5Y human neuroblastoma cells from hyperosmotic induced programmed cell death. J. Cell. Physiol. 1996; 166:323–31.

29. Miura M, Zhu H, Rotello R, Hartwieg EA and Yuan J. Induction of apoptosis in fibroblasts by IL-1 beta-converting enzyme, a mammalian homolog of the C. elegans cell death gene ced-3. Cell. 1993; 75:653–60.

30. Moule SK, and McGivan JD. Regulation of the plasma membrane potential in hepatocytes - mechanism and physiological significance. Biochim. Biophys. Acta. 1990; 1031:383–97.

31. Ohyama H, Yamada T, and Watanabe I. Cell volume reduction associated with interphase death in rat thymocytes. Radiation Res. 1981; 85:333–9.

32. Ohyama H, Yamada T, Ohkawa A, and Watanabe I. Radiation-induced formation of apoptotic bodies in rat thymus. Radiation Res. 1985; 101:123–30.

33. Orrenius S. Apoptosis: molecular mechanisms and implications for human disease. J. Intern. Med. 1995; 237:529–36.

34. Oyaizu N, and Pahwa S. Role of apoptosis in HIV disease pathogenesis. J. Clin. Immunol. 1995; 15:217–31.

35. Reed JC. Regulation of apoptosis by bcl-2 family proteins and its role in cancer and chemoresistance. Curr. Opin. Oncol. 1995; 7:541–6.

36. Roti-Roti LW, and Rothstein A. Adaptation of mouse leukemic cells (L5178Y) to anisotonic media. I. Cell volume regulation. Exp. Cell Res. 1973; 79:295–310.

37. Sanders EJ, and Wride MA. Programmed cell death in development. Int. Rev. Cytol. 1995; 163:105–73.

38. Schulte-Hermann R, Bursch W, Grasl-Kraupp B, Torok L, Ellinger A, Mullauer L. Role of active cell death (apoptosis) in multi-stage carcinogenesis. Toxicol. Lett. 1995; 82–83:143–8.

39. Schwartzman RA, and Cidlowski JA. Apoptosis: the biochemistry and molecular biology of programmed cell death. Endocrine Rev. 1993; 14:133–51.

40. Tata JR. Gene expression during metamorphosis: an ideal model for post-embryonic development. Bioessays 1993; 15:239–48.

41. Thomas N, and Bell PA. Glucocorticoid-induced cell-size changes and nuclear fragility in rat thymocytes. Mol. Cell. Endocrinol. 1981; 22:71–84.

42. Walev I, Reske K, Palmer M, Valeva A and Bhakdi S Potassium-inhibited processing of IL-1β in human monocytes. EMBO J. 1995; 14:1607–1614.

43. Wu G, and Flynn NE. Regulation of glutamine and glucose metabolism by cell volume in lymphocytes and macrophages. Biochim. Biophys. Acta. 1995; 1243:343–50.

44. Wyllie AH. Glucocorticoid-induced thymocyte apoptosis is associated with endogenous endonuclease activation. Nature 1980; 284:555–6.

45. Wyllie AH, and Morris RG. Hormone-induced cell death. Purification and properties of thymocytes undergoing apoptosis after glucocorticoid treatment. Am. J. Path. 1982; 109:78–87.

8

REGULATION OF p53-DEPENDENT APOPTOSIS IN HUMAN NEUROBLASTOMA CELLS BY ISOQUINOLINES

Victor C. Yu and Francesca Ronca

Institute of Molecular and Cell Biology
National University of Singapore
10 Kent Ridge Crescent
Singapore 119260, Republic of Singapore

1. ABSTRACT

We have studied staurosporine and a few other isoquinolines, which are known protein kinases inhibitors, for their ability in regulating the apoptosis in human neuroblastoma cells. Staurosporine and a subset of isoquinolinesulphonamides, including H-7 ([1-(5-Isoquinolinesulfonyl)-2-methylpiperazine]), were found to be able to induce widespread apoptosis, characterized by DNA fragmentation and nuclear condensation, in human neuroblastoma cells, SH-SY5Y, within 24 hours. Surprisingly, exposure of the cells to the H-7, but not staurosporine, caused a dramatic nuclear accumulation of p53. The kinetics of nuclear accumulation of p53 correlates well with the kinetics of induction of apoptosis. The effect of H-7 was further assessed in a group of human cell lines. Only cell lines harbouring the wild-type p53 gene were responsive to the stimulatory effect of H-7 on nuclear accumulation of p53. Furthermore, cell lines carrying a mutated p53 gene were resistant to the cytotoxic effect of H-7. The ability of the compound in mediating the apoptotic response in the SH-SY5Y line expressing a dominant negative mutant of p53 was significantly reduced. These data strongly suggest that a p53-dependent mechanism contributes to the cytoxicity of the H-7 in human neuroblastoma cells. Other PKC inhibitors failed to mediate apoptosis in neuroblastoma cells through the p53 pathway further suggest that a unique H-7 sensitive pathway which is different from the known PKC pathway is responsible for mediating the effect. Thus, the experimental paradigm of apoptosis triggered by H-7 in neuroblastoma cells is likely to be useful for gaining a further understanding of the pathways and mechanisms underlying the apoptotic function of p53.

2. INTRODUCTION

Apoptosis or programmed cell death was identified as a distinct process by virtue of the discrete series of morphological changes exhibited by cells undergoing programmed

Programmed Cell Death, edited by Shi *et al.*
Plenum Press, New York, 1997

cell death. Apoptotic cells experience viability loss accompanied by cytoplasmic blebbing, chromatin condensation, and DNA fragmentation (Wyllie *et al.*, 1980). The failure of cells to undergo apoptosis contributes to the origin and progression of human cancers.

The nuclear phosphoprotein p53 plays a key role in limiting the further expansion of cells containing damaged genomes. Loss of p53 functions in knock-out mice (Donehower *et al.*, 1992) results in tumour development early in life, whereas reconstitution of these functions in tumour cells usually confers growth arrest or apoptosis, depending on the cell type and circumstances (Yonish-Rouach, *et al.*, 1991; Harper *et al.*, 1993; El-Deiry *et al.*, 1993; Xiong *et al.*, 1993; Shaw *et al.*, 1992).

Over half of all human cancers are linked with loss of wild-type p53 function due to mutation of the *p53* gene (Vogelstein *et al.*, 1990; Hollstein *et al.*, 1991). If wild-type p53 function can also be lost as a result of epigenetic effect, the impact of this natural tumour suppressor may have been grossly underestimated. An understanding of the p53 response and its regulation in tumour cell lines carrying the wild-type *p53* gene are essential for evaluating this possibility.

The p53 protein can act as a transcription factor and its role in growth arrest by modulation of specific target genes is well documented (Harper *et al.*, 1993; El-Deiry *et al.*, 1993; Xiong *et al.*, 1993; Miyashita *et al.*, 1995; Owen-Schaub *et al.*, 1995). For example, the induction of the gene encoding the Cdk inhibitor, *p21/WAF1/CIP1*, by p53 results in G1 arrest of cells (Harper *et al.*, 1993; El-Deiry *et al.*, 1993; Xiong *et al.*, 1993). However, it is less certain how p53 regulates apoptosis. Some evidence indicates that induction of apoptosis by p53 may not require transcriptional activation of genes (Caelles *et al.*, 1994).

Neuroblastoma is one of the most common malignancies in childhood. It derives from the neural crest which gives rise to multiple cell lineages with neuronal, neurilemmal, or melanocytic phenotypes (Ross *et al.*, 1985; Sádee *et al.*, 1987). Unlike other tumour types, nearly all human neuroblastomas were found to carry wild-type *p53* gene (Vogen *et al.*, 1993; Komuro *et al.*, 1993; Imamura *et al.*, 1993; Hosoi *et al.*, 1994). Wild-type p53 protein is known to be present in many of the cell lines derived from neuroblastoma (Davidoff *et al.*, 1992). These cell lines are therefore suitable model systems for investigating the regulation and functions of wild-type p53 in tumour cells.

Staurosporine, a PKC inhibitor, was found to be an universal inducer of apoptosis in all mammalian cell lines tested, regardless of the state of differentiation and cell cycle phase. The broad apoptotic activity of staurosporine appears to be unique among other protein kinase inhibitors, raising the question whether a kinase mechanism is involved in the effect (Bertrand *et al.*, 1994). In this study, we are interested to study the apoptotic effect of a series of protein kinase inhibitors which are isoquinolinesulfonamide derivatives.

3. RESULTS AND DISCUSSION

3.1. A Subset of Isoquinolines Induced Apoptosis in Human Neuroblastoma Cells, SH-SY5Y

Upon treatment of cells with H-7 (50 μM) or staurosporine (100 nM) for 24 hours, the cells became rounded and loosely attached to the plate, suggesting the cells were losing viability. Further analyses of the cell samples documented that the cells were undergoing apoptotic death. Most of the morphological hallmarks associated with apoptosis were detectable, including internucleosomal fragmentation, which can be demonstrated by the DNA ladder assay (Fig. 1). The morphological changes resulted from H-7 and staurosporine treatment were indistinguishable (data not shown). Gross morphological changes asso-

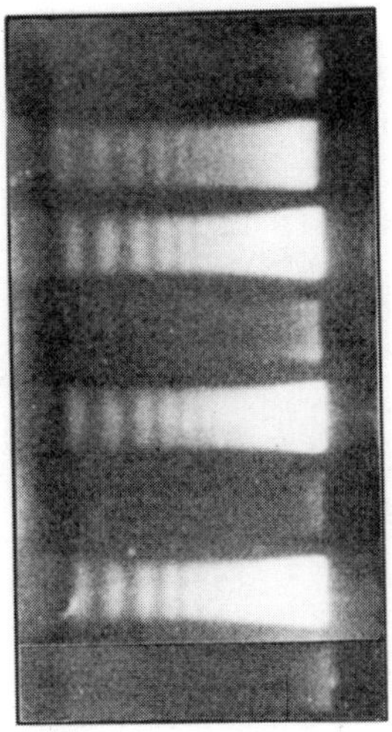

Figure 1. Induction of Apoptosis by H-7. Induction of apoptosis by protein kinase (PK) inhibitors was assessed by DNA ladder assay in human neuroblastoma cell line, SH-SY5Y. After 24 hours exposure to the PK inhibitors, cells were harvested and DNA extracted. The concentrations for all PK inhibitors used were 100 μM except for staurosporine which was 100 nM.

ciate with viability loss can be observed in 24 hours, whereas signs of apoptosis, detectable by DNA ladder assay, became apparent in 10–12 hours upon H-7 treatment. H-8 and H-9 were found to have similiar effect as H-7 on inducing dell death in these cells (Fig. 1). On the other hand, HA-1004, HA-1077 and iso-H-7, at concentrations up to 500 μM, did not induce noticeable morphological changes in these cells.

Upon H-7 treatment, the levels of nuclear p53 was found to be dramatically enhanced. Interestingly, no changes were noted in the levels of ICH-1_L, BCL-2, BCL-$x_{S/L}$, JUNB, c-FOS, RB, BAX and CDK-2, as a result of H-7 treatment (Ronca *et al.*, 1997).

Among the isoquinolines tested, there was a positive correlation between ability to induce apoptosis and ability to mediate nuclear accumulation of p53 (Fig. 2). It is of interest to note that despite the close structural resemblance of iso-H-7 to H-7, iso-H-7 is devoid of any inductive effects on apoptosis and nuclear accumulation of p53 (Fig. 2).

While Northern blot analysis revealed no significant changes in the *p53* mRNA levels (Ronca *et al.*, 1997), indicating that the nuclear accumulation of p53 was not due to an increase in *p53* transcripts, data from pulse-chase experiments revealed that H-7 had an effect in stabilizing the p53 in SH-SY5Y cells. The half-life of p53 was found to increase 10 fold, from 2.5 h to 28 h, upon H-7 treatment (Ronca *et al.*, 1997). The concentration range of H-7 (20–100 μM) which is required for the induction of apoptosis, measured by DNA fragmentation assay and by nuclear accumulation of p53, correlated well with each other (Ronca *et al.*, 1997).

At high concentration (20–100 μM), H-7 is not known to be a selective kinase inhibitor (Hidaka *et al.*, 1984). The reported activities of H-7 on kinase inhibition overlap with the other kinase inhibitors used in this study (Table 1). Staurosporine, HA-1004, HA-1077 and iso-H-7 had no effect on p53 induction even at high concentrations, raising the question of whether the effect of H-7 can be assigned to any well-established kinase pathway. It remains a formal possibility that the molecular target for H-7 in this case might not be a kinase.

3.2. H-7 Selectively Induced Apoptosis in Human Neuroblastoma Cells with Wild-Type p53

We further investigated the relationship between the induction of functional p53 and the subsequent apoptosis mediated by H-7. A panel of human neuroblastoma cell lines

Compound	Chemical structure	Induction of p53	Apoptosis
H-7	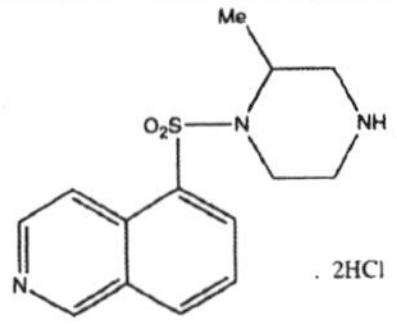	Yes	Yes
Iso-H-7	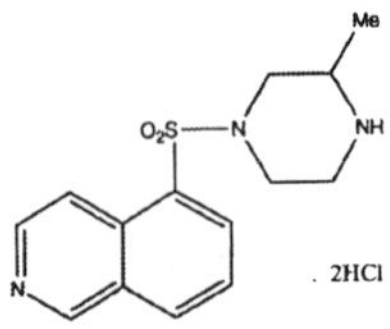	No	No
HA-1077	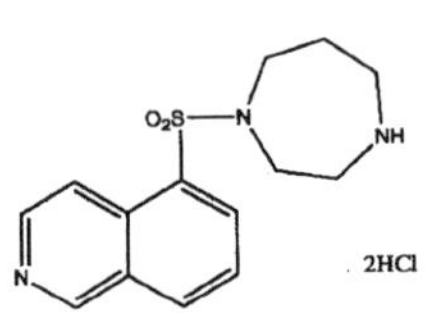	No	No
HA-1004	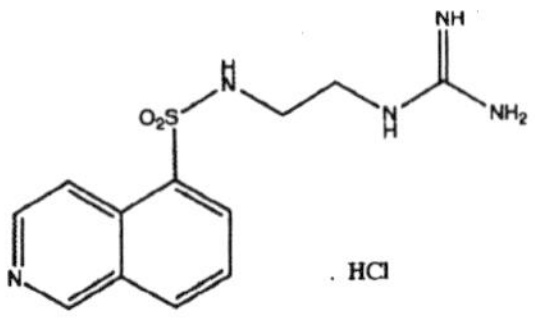	No	No
H-8	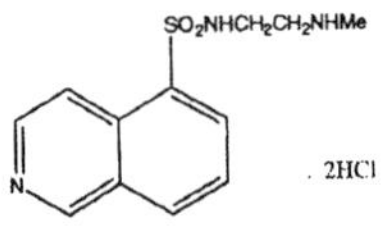	Yes	Yes
H-9	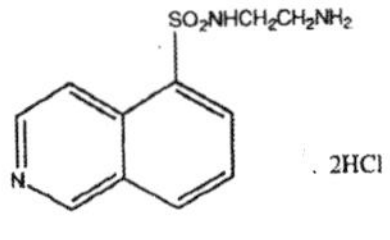	Yes	Yes
Staurosporine	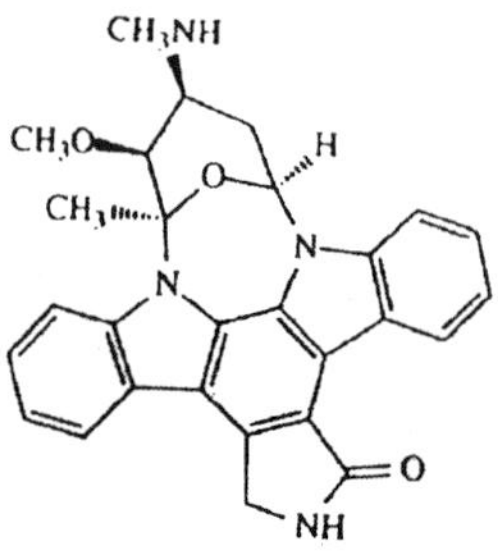	No	Yes

Figure 2. Effects of protein kinase inhibitors. Summary of the data on evaluating the effects of various protein kinase inhibitors on induction of nuclear p53 accumulation and apoptosis in SH-SY5Y human neuroblastoma cells.

Table 1. Summary of the K_i values of various protein kinase inhibitors
(Hidaka *et al.*, 1994; Seto *et al.*, 1991)

PK inhibitor	K_i				
	PKA	PKC	PKG	MLCK	CAMKII
H-7	3.0 μM	6.0 μM	5.8 μM	97 μM	
HA-1004	2.3 μM	40 μM	1.3 μM	150 μM	13 μM
Iso-H-7		50 μM (brain isoforms)			
HA-1077	1.6 μM	3.3 μM		20 nM	1.8 nM
Staurosporine	7.0 nM	0.7 nM	8.5 μM	1.3 nM	

carrying the wild-type *p53* gene were subjected to further analysis. These cells were sensitive to the cytotoxic and nuclear accumulation effects of H-7 (Fig. 3).

If the effect of H-7 on apoptosis requires wild-type p53, one would assume that H-7 would not be cytotoxic to many tumour lines expressing a mutated p53 protein. Indeed, H-7 failed to induce p53 and apoptosis in a panel of human cell lines with known mutations in the *p53* gene, including SW620, SW480, HT-29, T47D, HOS and Jeg-3 (Ronca *et al.*, 1997). Staurosporine, on the other hand, were equally effective in inducing apoptosis in all cell lines tested regardless of their p53 status (Ronca *et al.*, 1997).

3.3. Dominant Negative p53 Mutant Diminished the Cytotoxic Effect of H-7

In order to provide direct evidence that the cytotoxic effect of H-7 is correlated with the availability of functional p53 in the cell, we created a permanent SH-SY5Y line expressing the dominant negative mutant of p53, tagged with a 9 amino acid peptide of influenza hemoagglutinin (HA) (Fig. 4).

This mini protein contains the C-terminal portion of p53 (amino acids 302–393). This C-terminal domain of p53, when expressed separately, exhibits strong transforming activity in cells which normally express low levels of endogenous p53 (Shaulian *et al.*, 1995). This transforming activity is attributable, at least in part, to a negative dominant mechanism, involving the formation of non-functional hetero-oligomers between the C-

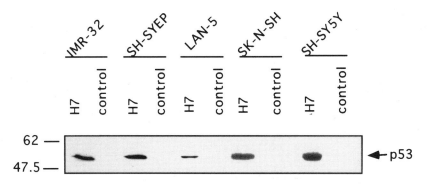

Figure 3. The effect of H-7 on nuclear accumulation of p53 protein in human neuroblastoma cell lines. Immunoblot analyses of nuclear p53 were performed to examine the effect of H-7 on p53 induction in these cells. Equal amounts of nuclear extracts from each indicated cell line before and after treatment with 50 μM of H-7 for 10–16 h were used for the analysis.

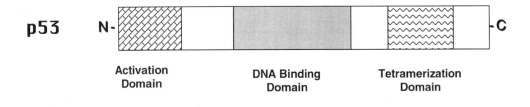

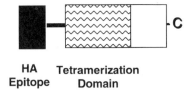

Figure 4. The dominant-negative mutant of p53. A cell line transfected with the hygromycin selection vector alone was also selected and served as the vector control line. The line expressing the dominant p53 negative mutant was created and was named, SY5.6.

terminal fragments and the intact endogenous wild-type molecule (Shaulian *et al.*, 1995; Shaulian *et al.*, 1992). The dominant negative p53 mutant cell line, named SY 5.6, was found to express the p53 mini protein as detected by immunoblot analysis using the rabbit anti-HA polyclonal antibody (Ronca *et al.*, 1997). While the ability of staurosporine in mediating apoptosis was not affected in line SY 5.6 in comparison to vector control line, H-7-mediated cell killing had clearly become less effective in cells expressing the dominant negative mutant of p53 (Table 2).

4. CONCLUSIONS

The compound H-7 selectively induces apoptosis in human neuroblastoma cells. Several lines of evidence suggested that the apoptotic effect of H-7 in these cells was mediated, at least in part, through p53. 1. The kinetics of induction of apoptosis and p53 ac-

Table 2. Comparison of the effect of H-7 and staurosporine on induction of apoptosis in vector control versus the dominant negative line, SY5.6. Cells from both lines were grown in the presence of hygromycin. The cell lines were plated onto a 96 well-plates and the viability of the cells was measured by MTT assay

	H-7 (μM)		Staurosporine (nM)	
Concentration	Vector	Sy 5.6	Vector	Sy 5.6
0	100 ± 1.14	100 ± 1.53	100 ± 1.72	100 ± 5.97
30	77.20 ± 0.37	94.41 ± 0.77	74.93 ± 3.97	77.99 ± 5.99
40	67.68 ± 2.30	80.28 ± 2.63	N.D.	N.D.
50	47.63 ± 2.38	70.71 ± 4.12	63.31 ± 7.96	71.69 ± 2.39
70	23.53 ± 1.70	70.85 ± 1.61	19.05 ± 3.84	12.43 ± 7.61
100	23.10 ± 3.32	57.72 ± 3.66	13.66 ± 5.10	9.76 ± 3.91

N.D. = not determined

cumulation were similar. 2. Isoquinoline analogues that failed to induce apoptosis were also ineffective in p53 induction. 3. In cell lines expressing a mutant p53, H-7 failed to induce apoptosis. 4. Expression of a dominant-negative form of p53 was sufficient to counteract the cytotoxic effect of H-7. A subset of isoquinolinesulphonamide derivatives appear to possess a unique activity in regulating p53 which is not shared among many other PKC inhibitors.

ACKNOWLEDGMENTS

We are grateful to Drs. Edward Harlow and Steven Elledge for human p53 and p21 cDNA, respectively; and to Drs. Wolfang Sádee and R.C. Seeger for human neuroblastoma cell lines. This work was supported by grants from the National Science and Technology Board of Singapore.

REFERENCES

Bertrand, R., Solary, E., O'Connor, P., Kohn, K.W., and Pommier, Y. (1994) *Exp. Cell Res.* **211**, 314–321

Caelles, C., Helmberg, A., and Karin, M. (1994) *Nature* **370**, 220–223

Davidoff, A.M., Pence, J.C., Shorter, N.A., Iglehart, J.D., and Marks, J.R. (1992) *Oncogene* **7**, 127–133

Dole, M.G., Jasty, R., Cooper, M.J., Thompson, C.B., Nunez, G., and Castle, V.P. (1995) *Cancer Res.* **55**, 2576–2582

Donehower, L.A., Harvey, M., Slagle, B.L., McArthur, M.J., Montgomery, C.A.,Jr., Butel, J.S., and Bradley, A. (1992) *Nature* **356**, 215–221

El-Deiry, W.S., Tokino, T., Velculescu, V.E., Levy, D.B., Parsons, R., Trent, J.M., Lin, D., Mercer, W.E., Kinzler, K.W., and Vogelstein, B. (1993) *Cell* **75**, 817–825

Harper, J.W., Adami, G.R., Wei, N., Keyomarsi, K., and Elledge, S.J. (1993) *Cell* **75**, 805–816

Hidaka, H., Inagaki, M., Kawamoto, S., and Sasaki, Y. (1984) *Biochemistry* **23**, 5036–5041

Hollstein, M., Sidransky, D., Vogelstein, B., and Harris, C.C. (1991) *Science* **253**, 49–53

Hosoi, G., Hara, J., Okamura, T., Osugi, Y., Ishihara, S., Fukuzawa, M., Okada, A., Okada, S., and Tawa, A. (1994) *Cancer* **73**, 3087–3093

Imamura, J., Bartram, C.R., Berthold, F., Harms, D., Nakamura, H., and Koeffler, H.P. (1993) *Cancer Res.* **53**, 4053–4058

Komuro, H., Hayashi, Y., Kawamura, M., Hayashi, K., Kaneko, Y., Kamoshita, S., Hanada, R., Yamamoto, K., Hongo, T., and Yamada, M. (1993) *Cancer Res.* **53**, 5284–5288

Miyashita, T. and Reed, J.C. (1995) *Cell* **80**, 293–299

Owen-Schaub, L.B., Zhang, W., Cusack, J.C., Angelo, L.S., Santee, S.M., Fujiwara, T., Roth, J.A., Deisseroth, A.B., Zhang, W.W., and Kruzel, E. (1995) *Mol. Cell Biol.* **1555**, 3032–3040

Ronca, F., Chan, S.L., and Yu, V.C. (1997) *J. Biol. Chem.*, **272**, 4252–4260

Ross, R.A. and Biedler, J.L. (1985) *Cancer Res.* **45**, 1628–1632

Sádee, W., Yu, V.C., Richards, M.L., Preis, P.N., Schwab, M.R., Brodsky, F.M., and Biedler, J.L. (1987) *Cancer Res.* **47**, 5207–5212

Seto, M., Sasaki, Y., Sasaki, Y., and Hidaka, H. (1991) *Eur. J. Pharmacol.* **195**, 267–272

Shaulian, E., Zauberman, A., Ginsberg, D., and Oren, M. (1992) *Mol. Cell Biol.* **12**, 5581–5592

Shaulian, E., Haviv, I., Shaul, Y., and Oren, M. (1995) *Oncogene* **10**, 671–680

Shaw, P., Bovey, R., Tardy, S., Sahli, R., Sordat, B., and Costa, J. (1992) *Proc. Natl. Acad. Sci. U. S. A.* **89**, 4495–4499

Vogan, K., Bernstein, M., Leclerc, J.M., Brisson, L., Brossard, J., Brodeur, G.M., Pelletier, J., and Gros, P. (1993) *Cancer Res.* **53**, 5269–5273

Vogelstein, B. (1990) *Nature* **348**, 681–682

Wyllie, A.H., Kerr, J.F., and Currie, A.R. (1980) *Int. Rev. Cytol.* **68**, 251–306

Xiong, Y., Hannon, G.J., Zhang, H., Casso, D., Kobayashi, R., and Beach, D. (1993) *Nature* **366**, 701–704

Yonish-Rouach, E., Resnitzky, D., Lotem, J., Sachs, L., Kimchi, A., and Oren, M. (1991) *Nature* **352**, 345–347

9

INDUCIBLE Fas-RESISTANCE IN B LYMPHOCYTES

Thomas L. Rothstein,[1,2,4] Thomas J. Schneider,[2] Ann Marshak-Rothstein,[2,3] and Linda C. Foote[2]

[1]Department of Medicine
[2]Department of Microbiology
[3]Department of Pathology
[4]Evans Memorial Department of Clinical Research
Boston University Medical Center
Boston, Massachusetts 002118

ABSTRACT

Activation of B lymphocytes, such as that provided by CD40L, is required for induction of susceptibility to Fas-mediated apoptosis. However, all forms of B cell activation do not promote Fas-sensitivity. Anti-immunoglobulin antibody, and IL-4, produce a state of Fas-resistance in otherwise sensitive (CD40L-stimulated) B cell targets through at least partially distinct signaling pathways. Inducible Fas-resistance may act to thwart B cell deletion mediated by FasL-expressing CD4[+] Th1 cells, either prior to or during germinal center formation. This may promote the viability of autoreactive B cells, thereby fostering autoantibody formation as observed in IL-4 overexpressing transgenic mice, with implications for normal- and auto-immunity.

INTRODUCTION

As important as proliferation is to the early development and later expansion of lymphocyte effector cells, self-tolerance, antigen specificity, and the timely termination of immune responses depend on cell death. Cell death programs are initiated in lymphocytes through several means, including cytokine growth factor withdrawal and specific receptor triggering. One receptor that plays a key role in producing lymphocyte cell death is CD95 (Fas, APO-1), a 48 kDa transmembrane glycoprotein that is a member of the extended TNFR family[1]. The Fas receptor triggers apoptosis following engagement by its cognate ligand, Fas ligand (FasL), a cell surface molecule that is found primarily on activated CD4[+] Th1 and CD8[+] T lymphocytes[2]. Fas is expressed by cells of hematopoietic origin, as well as nonhematopoietic tissues including liver, ovary, and heart[3]. Fas mutations that dis-

Programmed Cell Death, edited by Shi *et al.*
Plenum Press, New York, 1997

rupt function, as found in *lpr/lpr* mice and in patients with ALPS (autoimmune lympho-proliferation syndrome), are associated with the expression of autoantibodies, implicating Fas in the regulation of autoreactive B cells[4-6]; because bone marrow function appears normal in Fas-deficient animals, Fas may regulate lymphocyte numbers predominantly in the periphery[1].

Fas triggering for cell death is controlled by inducible Fas expression in activated lymphocytes, and by alteration of the target cell level of susceptibility to Fas signaling, both of which we have found to be regulated in a receptor-specific fashion[7-10]. Thus, stimulation of primary B lymphocytes through the CD40 receptor, another member of the TNFR family, produces exquisite sensitivity to Fas-mediated apoptosis, whereas stimulation through the antigen receptor (surface immunoglobulin or sIg) fails to do so, even though each receptor is individually capable of mediating a complete mitogenic response[10]. Moreover, the effect of sIg signaling is dominant, such that B cells stimulated by both anti-Ig (a surrogate for T-independent type II antigens) and CD40L resist Fas-mediated apoptosis[10]. Thus, engagement of the B cell antigen receptor produces a state of Fas-resistance in otherwise sensitive targets in which Fas signaling for cell death is suppressed. This has now been shown to be true of both murine and human B cells, and indicates that such cells are not simply passive targets for Fas killing, but rather pro-actively alter their endogenous level of susceptibility to Fas-mediated apoptosis in a receptor-specific fashion as a response to environmental cues[10,11].

Recently we have demonstrated that sIg-induced Fas-resistance results from an intrinsic or intracellular change in B cell targets rather than an alteration in cell surface interaction molecules[12]. Three lines of evidence support this conclusion: 1) Fas expression that is upregulated by CD40L remains elevated when B cells are stimulated by the combination of anti-Ig plus CD40L, dual treatment that induces Fas-resistance, ruling out diminished Fas expression as an explanation for loss of Fas-sensitivity[10]; 2) Unlabeled dual treated (Fas-resistant) B cells compete for Th1 cell-mediated killing of labeled CD40L-stimulated (Fas-sensitive) B cell targets as well as unlabeled CD40L-stimulated (Fas-sensitive) B cells ("cold competition" analysis), ruling out an alteration in either Fas conformation or the expression of other cell surface molecules that might be required for T cell-B cell interaction[12]; and, 3) B cell targets that are sensitive (CD40L-stimulated) and resistant (anti-Ig plus CD40L-stimulated) to apoptosis induced by FasL-expressing Th1 cells are similarly sensitive and resistant to apoptosis induced by recombinant FasL presented as either a soluble protein or as purified insect cell membranes (kindly provided by Drs. David Lynch, Immunex, Seattle, WA and Marilyn Kehry, Boehringer-Ingelheim, Ridgefield, CT, respectively), ruling out participation by other receptor-ligand pairs in Th1-mediated apoptosis[12].

With this background we have now explored additional features of B cell resistance to Fas-mediated apoptosis, including mechanisms that might underlie Fas-resistance, cytokine triggers that elicit Fas-resistance, and the relationship between Fas-resistance and B cell anergy and autoantibody formation.

RESULTS

Durability and Stability of sIg-Mediated Fas-Resistance. Inducible Fas-resistance might be expected to play a significant role in immune responsiveness to the extent that it represents a durable and stable change in functional phenotype, and, conversely, the role of Fas-resistance may be very limited if induction is only transient. The durability of Fas-resistance was tested by stimulating primary murine B cells with either CD40L (to produce Fas-sensitivity) or the combination of anti-Ig plus CD40L (to produce Fas-resistance) for various periods of time after which B cells were tested as targets for Th1 cell-mediated cytotoxicity (Th1-CMC) in standard 4 hour lectin dependent ^{51}Cr release assays using AE7 effectors that have previously been shown to kill in a Fas-dependent fashion[8,10]. CD40L was used as a soluble, chimeric protein in which the extracellular domain of murine CD40L is fused to the extracellular domain of CD8α[13]; soluble protein secreted by the recombinant J558L cell line (kindly provided by Dr. Peter Lane) was purified as previously described and in each experiment crosslinked *in situ* with anti-CD8[14]. B cells stimulated with crosslinked CD40L for as little as 18 hours expressed elevated levels of Fas as detected by immunofluorescent staining and were sensitive to Fas-mediated apoptosis induced by Th1 effectors (specific lysis, 40–50%). Similar levels of Fas-mediated apoptosis were found when the same B cells were treated with CD40L for 42 hours and for 66 hours prior to assessment of Th1-CMC. Anti-Ig-induced Fas-resistance followed a comparable pattern. Resistance to Th1-CMC was well established when B cells were stimulated with the combination of anti-Ig plus CD40L for as little as 18 hours (specific lysis, 0–10%), and Fas-resistance was clearly evident after dual treatment for 42 hours and 66 hours, although the magnitude of Fas-resistance declined a little at 66 hours. Thus, Fas-resistance is induced by anti-Ig over a broad time scale and is not dependent on a restricted set of culture conditions.

Along the same lines, the stability of Fas-resistance was tested by stimulating B cells with either CD40L (to produce Fas-sensitivity) or the combination of anti-Ig plus CD40L (to produce Fas-resistance) for 24 hours, after which B cells were washed, incubated in medium for an additional 24 hours, and tested as targets for Th1-CMC. These B cell targets were found to be as sensitive and resistant to Fas-mediated apoptosis, respectively, as B cells treated without interruption for 48 hours with either CD40L alone or the combination of anti-Ig plus CD40L (Figure 1). Thus, induction of Fas-sensitivity and Fas-resistance represent stable changes in the functional phenotype of B cell targets.

Mechanism of Fas-Resistance. Complete Fas-resistance is induced in primary B cells when anti-Ig is present during the last 12–24 hours of 48 hour cultures with CD40L, and substantial (more than half-maximal) Fas-resistance is established by anti-Ig treatment for just 6 hours, although there is no resistance produced by adding anti-Ig at 0 hours (at

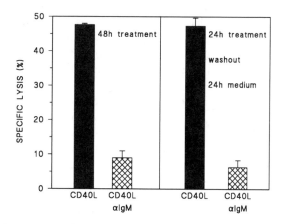

Figure 1. Inducible Fas-resistance is stable. Purified splenic B cells from BALB/cByJ mice were stimulated for 48 hours with either CD40L alone, or the combination of anti-IgM plus CD40L, as indicated (left hand panel) or were stimulated for 24 hours, washed, and then cultured in medium for an additional 24 hours (right hand panel). B cells were then tested for susceptibility to Th1-mediated, Fas-dependent cytotoxicity in standard 4 hour lectin-dependent ^{51}Cr release assays using AE7 CD4[+] Th1 effectors at an effector:target cell ratio of 9:1, as previously described[10]. For each condition, mean percent specific lysis of triplicate assays is shown, along with a line indicating the standard error of the mean. Reagents used were: CD40L, dialyzed supernatant from J558L cells that secrete a CD40L-CD8α fusion protein[13] at 1:10, plus supernatant from the 53–6–72 hybridoma containing anti-CD8α antibody at 1:80, as described[10,12]; αIgM, F(ab')[2] fragments of goat anti-mouse IgM antibody at 10 μg/ml.

the start of ^{51}Cr release assays) demonstrating that some time is required[10,12]. The latter observation raises the possibility that induction of Fas-resistance depends on synthesis of a new molecule, rather than posttranslational modification of a pre-existing one. Using the shorter, 6 hour period of treatment we have shown that induction of Fas-resistance is interrupted by addition of either cycloheximide[12] or actinomycin D concurrently with anti-Ig, and thus in B cells resistance against Fas-mediated apoptosis depends on new protein synthesis and, in turn, new gene transcription. Much effort has been focused on identifying the nature of the protein whose expression confers Fas-resistance.

It has been suggested that Fas-mediated apoptosis depends on the production of ceramide by sphingomyelinase, and certain ceramide analogs are capable of penetrating cell membranes and inducing apoptosis[15–17]. One approach to focusing efforts for protein identification would be to determine whether suppression of Fas signals occurs early (eg, interruption of tyrosine phosphorylation) or late (eg, resistance against ceramide-mediated apoptosis). To evaluate this, B cells were stimulated with either CD40L alone or the combination of anti-Ig plus CD40L and tested for susceptibility to apoptosis in ^{51}Cr release assays in which effector cells were replaced by apoptotic C2 ceramide or the control, non-apoptotic analog, C2 dihydroceramide. CD40L-stimulated, Fas-sensitive B cells were found to be susceptible to cytotoxicity mediated by C2 ceramide, whereas dual treated (anti-Ig plus CD40L), Fas-resistant targets were not. As expected, neither Fas-sensitive nor Fas-resistant B cells were killed by the control, C2 dihyrdoceramide. Thus, resistance to Th1- and rFasL-mediated apoptosis is associated with resistance to cytotoxicity produced by a ceramide analog, suggesting that suppression of signaling for apoptosis acts relatively late in the cell death pathway.

Prominent among the molecules associated with broad-based suppression of apoptosis are members of the Bcl-2 family of proteins[18]. We assessed gene and protein expression of a number of Bcl-2 homologs and interacting proteins, comparing B cells stimulated either with CD40L alone or with anti-Ig during the last 0–48 hours of a 48 hour culture period with CD40L. Specific induction of *bcl-x*$_L$ mRNA[19] (by quantitative RT-PCR) and Bcl-x$_L$ protein (by Western blot) was observed in dual treated B cells. This raises the possibility that Bcl-x$_L$ is responsible, at least in part, for B cell Fas-resistance; however, the timing of Western blot detectable Bcl-x$_L$ protein appears to be somewhat delayed in comparison to the onset of Fas-resistance, suggesting that if Bcl-x$_L$ plays a role in producing Fas-resistance, other proteins may collaborate in establishing this phenotype, particularly at early time points.

In order to identify additional gene transcripts that contribute to Fas-resistance, we have utilized differential display[20], comparing mRNA obtained from B cells stimulated with CD40L for 48 hours with mRNA obtained from B cells stimulated with anti-Ig for the last 6 hours of 48 hour cultures with CD40L. Differential display provides the opportunity to identify genes specifically expressed under conditions that produce Fas-resistance without any information as to their likely sequence or function. A number of putative transcripts have been identified in this way, and thus far 3 have been confirmed by Northern blotting, each within the range of 2–4 kb. These fragments are in the process of being subcloned and sequenced; at present their identity and relationship to Fas-resistance is unknown, but they present the possibility that in this way additional, perhaps novel, genes responsible for modulating susceptibility to Fas-mediated apoptosis will be found.

IL-4 Induced Fas-Resistance. A number of lymphokines that influence proliferation and/or differentiation of B cells have been tested for their effect on susceptibility to Fas-mediated apoptosis, including IL-2, IL-4, IL-5, IL-6, IFNγ and TNFα. Among these, only

IL-4 substantially altered the Fas-sensitivity of CD40L-treated B cells, inducing Fas-resistance in otherwise sensitive targets, as previously observed with anti-Ig. However, IL-4 at optimal doses (100 units/ml) induced resistance against Fas-mediated apoptosis that was not quite as complete as that produced by anti-Ig. In general terms, IL-4 plus CD40L-treated B cell targets required an approximately 10-fold higher effector:target cell ratio to produce levels of specific cell lysis comparable to that observed with B cells treated with CD40L alone, whereas anti-Ig plus CD40L-treated B cell targets required an approximately 20-fold higher effector:target cell ratio. IL-4-induced Fas-resistance was observed with recombinant IL-4, and was reversed with 11B11 neutralizing anti-IL-4 antibody[21], demonstrating specificity for IL-4. IL-4-induced Fas-resistance was not associated with a change in the elevated level of surface Fas expression produced by CD40L, and IL-4 Fas-resistant B cells resisted the apoptotic effect of recombinant FasL; taken together these results suggest that IL-4, like anti-Ig, suppresses Fas killing through an intracellular rather than an extracellular change in B cell targets.

Distinct Pathways for Fas-Resistance Induced by IL-4 and Anti-Ig. In general, IL-4 and anti-Ig appear to act through signaling pathways that are for the most part distinct[22,23], although recently both IL-4 and anti-Ig have been shown to stimulate STAT6 activation[24,25]. To determine whether IL-4 and anti-Ig trigger the same pathway leading to Fas-resistance, the time course for induction of resistance to Fas killing was determined, by adding either reagent at various times prior to the end of 48 hour B cell cultures with CD40L. In a series of experiments, anti-Ig induced maximal Fas-resistance within 12 hours, whereas IL-4 did not induce maximal Fas-resistance until 24 hours (Figure 2). This was especially evident when B cells were exposed for 6 hours, by which time anti-Ig had produced more than half-maximal resistance whereas IL-4 had produced little change in Fas-sensitivity. Thus, the rate at which resistance to Fas killing develops distinguishes IL-4 and anti-Ig as agents producing Fas-resistance.

As a second test for signaling differences, the role of PKC in bringing about Fas-resistance was examined by evaluating Th1-CMC of B cell targets treated with either IL-4 or anti-Ig after PKC depletion by incubation with the phorbol ester PKC agonist, phorbol myristate acetate (PMA)[26]. As a control, B cells were treated with IL-4 or anti-Ig after incubation with the inactive phorbol ester analog, 4αPMA. Experiments were carried out by

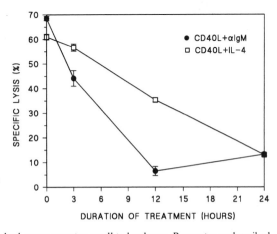

Figure 2. Fas-resistance is induced by anti-IgM and by IL-4 with distinct time courses. Purified splenic B cells from BALB/cByJ mice were stimulated for 48 hours with CD40L alone (0 hours), or with the combination of anti-IgM plus CD40L (closed circles) or IL-4 plus CD40L (open squares), in which anti-IgM or IL-4 was added at the indicated number of hours prior to the end of 48 hour cultures. B cells were then tested for susceptibility to Th1-mediated, Fas-dependent cytotoxicity in standard 4 hour lectin-dependent ^{51}Cr release assays as described in the legend to Figure 1. Mean percent specific lysis of triplicate assays is shown as a function of duration of treatment before the end of culture for anti-IgM and for Il-4, along with a line indicating the standard error of the mean. In some cases standard errors were too small to be shown. Reagents are described in the legend to Figure 1, except for IL-4, which was used as the recombinant gene product at 100 units/ml.

incubating B cells for a full 48 hours concurrently with CD40L plus PMA (or 4αPMA) for 48 hours, and then adding IL-4 or anti-Ig 24 hours before the end of culture. Remarkably, the presence of PMA (or 4αPMA) did not alter the induction of Fas-sensitivity in B cells stimulated by CD40L alone in any way (consistent with the PKC-independence of CD40 signaling[14,27]). However, prior depletion of PKC blocked induction of Fas-resistance by anti-Ig; in contrast, there was no difference in Fas-resistance of B cells treated with IL-4 in the presence of PMA as opposed to 4αPMA. Thus, induction of Fas-resistance by IL-4 is PKC-independent, whereas induction by anti-Ig is PKC-dependent. PKC-dependence represents a second criterion by which Fas-resistance induced by IL-4 and anti-Ig are distinguished, and further suggests that these agents trigger Fas-resistance through at least partially non-overlapping intracellular signaling pathways.

Synergy between IL-4 and Anti-Ig. The differences in signaling for IL-4- and anti-Ig-induced Fas-resistance suggested that these agents might act to complement each other. To test this possibility, B cells were treated with suboptimal doses of IL-4 and anti-Ig, either alone or together, for the last 24 hours of 48 hour cultures with CD40L, after which Th1-CMC was assessed. Alone, these reagents produced less than half-maximal resistance to Fas apoptosis, but together, Fas-resistance was complete. This result suggests that sIg and IL-4R triggered events can act in concert to produce full resistance against Fas-mediated apoptosis at lower doses of the corresponding ligands when present together, as opposed to separately.

Inducible Fas-Resistance in Anergic B Cells. Fas-deficient *lpr/lpr* mice manifest extensive autoantibody production[4]. It may be that inducible Fas-resistance, like genetic Fas-deficiency, regulates the ability of autoreactive B cells to escape deletional pathways and become part of the immunocompetent pool. To begin to examine this issue, we determined whether IL-4 is capable of inducing Fas-resistance in anergic B cells (noting that sIg signaling has been shown to be defective in anergic B cells, and is thus likely incapable of mediating Fas-resistance[28,29]). The anti-HEL/HEL double transgenic mouse model developed by Goodnow and colleagues was used[30]. In this model, mice express transgenes for a B cell antigen receptor that recognizes hen egg lysozyme (HEL), and for soluble HEL. B cells that develop in the continual presence of soluble autoantigen become anergic. These anergic B cells are present in the periphery in near normal numbers; however, their lifespan is shortened to several days, subsequent deletion occurs through T cell induced, Fas-mediated apoptosis, sIg expression is reduced, and, as noted earlier, sIg signaling is defective[28-33]. B cells from double transgenic (anti-HEL/HEL) mice were stimulated with CD40L alone (to produce Fas-sensitivity), and with the combination of IL-4 plus CD40L (to produce Fas-resistance), for 16 hours, and then tested for susceptibility to Th1-CMC. Results were compared with single transgenic (anti-HEL) mice that express HEL-specific antigen receptors but are not tolerized because B cells are not exposed to cognate antigen during development. In both cases, treatment with CD40L led to similar levels of specific lysis following exposure of B cell targets to FasL-expressing Th1 cells; furthermore, in both cases IL-4 produced similar levels of Fas-resistance (Figure 3) without any change in the CD40L-upregulated level of surface Fas expression. Thus, Fas-resistance is induced by IL-4 in anergic, autoreactive B cells, and, at least as far as CD40 and IL-4R are concerned, anergic B cells express the same receptor-specific regulation of susceptibility to Fas-mediated apoptosis as conventional B cells.

Autoantibodies in IL-4 Overexpressing Transgenic Mice. Because Fas-resistance is induced in anergic B cells by IL-4, it might be speculated that elevated expression of IL-4

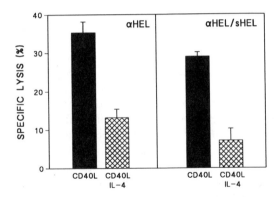

Figure 3. IL-4 induces Fas-resistance in anergic B cells. Purified splenic B cells were prepared from anti-HEL/soluble HEL (αHEL/sHEL) double transgenic mice, in which B cells are anergic, and from αHEL single transgenic mice, in which B cells function normally. The genetic status of mice used in these experiments was confirmed by PCR amplification of DNA. B cells were stimulated for 16 hours with CD40L alone (solid bars), or with the combination of IL-4 plus CD40L (hatched bars). B cells were then tested for susceptibility to Th1-mediated, Fas-dependent cytotoxicity in standard 4 hour lectin-dependent ^{51}Cr release assays as described in the legend to Figure 1. The mean of results from three separate experiments is shown for each condition, along with a line indicating the standard error of the mean.

would insulate autoreactive B cells against Fas-mediated deletion and provide the opportunity for differentiation and autoantibody formation, as is observed in Fas-deficient mice. To test this we examined transgenic mice that overexpress IL-4 (TG.UD line), originally developed by Philip Leder and colleagues[34], and kindly provided by Drs. Leder and Abul Abbas (Harvard Medical School, Boston, MA). Sera from these mice were tested for the presence of autoantibodies by an indirect immunofluorescence technique using fixed HEp-2 cells[35]. Five out of 5 IL-4 transgenic sera (and no normal sera) expressed substantial levels of autoantibody that appeared to recognize cytoplasmically located intracellular components. Thus, inducible Fas-resistance appears to alter the normal disposition of autoreactive B cells in these mice.

DISCUSSION

The work presented here, as well as additional studies from this laboratory and from other investigators, demonstrates that B cells are not simply passive targets for Th1-induced apoptosis, but rather regulate their own susceptibility to Fas-directed cell death in a receptor-specific fashion in which antigen and Th2-derived IL-4 function to limit the deleterious effects of Fas signaling. However, differences in time course and PKC-dependence between Fas-resistance induced by anti-Ig and IL-4 suggests that distinct intracellular signaling pathways are involved.

Engagement of CD40, sIg and IL-4R produce differing effects on B cell fate depending on the nature of the system under study[36]. Thus, in immature B cells, CD40 and IL-4R rescue from sIg-induced apoptosis; in germinal center B cells, both sIg and CD40 counteract spontaneous apoptosis, while IL-4R has no effect; and, as noted above, in mature B cells CD40 promotes, whereas sIg and IL-4R oppose, Fas-mediated apoptosis. Thus, signals generated by specific receptor engagement may regulate viability differently depending on the developmental state of the B cell and/or the trigger for apoptosis. Of interest, a recent report indicates that Fas-resistant T cells overexpress Th2 cytokines, specifically including IL-4[37].

Once established, Fas-sensitivity and Fas-resistance appear to be relatively durable and stable. This contrasts with a recent report in which Fas-resistance was more apparent at 2 days than at 1 day or 3 days[38]. However, in this report the level of Fas-resistance was meager, and apoptosis was induced by the Fas-specific Jo-2 monoclonal antibody[39]. We have previously shown that Jo-2 induces apoptosis in dual stimulated B cells that resist cytotoxicity mediated by Th1 effector cells and by recombinant FasL, indicating that Jo-2 produces a non-physiological or super-physiological signal[12]. This is likely due to hyper-crosslinking of the Fas-specific antibody by B cell FcR, and is blocked by the FcR-specific 2.4G2 monoclonal antibody[12], but could also relate to binding of a different determinant by Jo-2 (unpublished observations).

The specific mechanisms responsible for inducible Fas-resistance remain uncertain. Fas-resistance induced by anti-Ig depends on transcriptional activation and counteracts apoptosis produced by ceramide, suggesting that a newly synthesized protein interferes with Fas signaling at a point beyond initial signaling events. A candidate would be $Bcl-x_L$, which is induced by anti-Ig in CD40L stimulated B cells. Further, in very preliminary experiments in collaboration with Dr. Gabriel Nunez (University of Michigan, Ann Arbor, MI), transgenic mice overexpressing $bcl-x_L$ fail to become sensitive to Fas-mediated apoptosis following stimulation by CD40L despite upregulation of Fas expression. However, these results do not conclusively implicate $bcl-x_L$ in Fas-resistance and the late appearance of $Bcl-x_L$ on Western blot suggests the possibility of discordance with the earlier onset of partial Fas-resistance (although this may relate to the relative sensitivity of the two assays). In view of this we have undertaken to identify and isolate genes associated with induction of Fas-resistance by differential display of variously treated primary B cell populations. Thus far 3 differentially expressed bands have been confirmed as candidate genes by Northern blotting. Future work will focus on cloning and sequencing these candidate genes, and further evaluation of their role in, and regulation during, inducible Fas-resistance. More and more information regarding the mechanisms by which Fas triggering leads to cell death is becoming available, including information on Fas-associated proteins, proteins that associate with Fas-associated proteins, metabolic changes following Fas signaling, and the role of ICE-like proteases[20,21,40]. One or more of these steps may represent a control point for suppression of Fas signaling, and the nature of differentially expressed genes may rapidly clarify the level at which Fas-resistance is established; alternatively, multiple mechanisms and multiple gene products may be involved.

The twin observations that IL-4 induces Fas-resistance in anergic B cells and that IL-4 transgenic mice express autoantibodies suggest that inducible resistance to Fas-mediated apoptosis may regulate the disposition of autoreactive B cells. Autoreactive B cells express receptors with forbidden specificities, and several mechanisms exist for their exclusion from the immunocompetent pool[41]. These include deletion in the bone marrow, receptor editing, and anergy, dictated, at least in part, by the avidity of the interaction between sIg and autologous antigen. In murine models, anergic B cells are briefly present in the periphery in a state of diminished responsiveness with respect to antigen receptor signaling, although responses to CD40L and IL-4 remain intact[28,29,42]; anergic B cells are subsequently deleted through a Fas-dependent mechanism[33]. In normal individuals, anergic, autoreactive B cells appear to be present in peripheral lymphoid tissues on the basis of phenotypic parallels with B cells from double transgenic murine models[43]. Thus, interference with the Fas apoptotic pathway in normal individuals may interfere with deletion of anergic, autoreactive B cells, thereby providing the opportunity for such B cells to join the immunocompetent pool, as may occur in *lpr/lpr* mice in which Fas-resistance is established on the basis of genetic deficiency in Fas expression. In the *lpr* model, autoantibody producing B cells are found within the T cell rich areas of the periarteriolar lymphoid sheath (PALS), suggesting that Fas killing regulates autoreactive B cell survival at a stage prior to germinal center formation[44]. Inducible Fas-resistance may regulate anergic B cell survival at the same stage, and in this context experiments in progress will reveal whether the autoantibody producing B cells in IL-4 transgenic mice also localize to the inner PALS.

The regulation of anergic, autoreactive B cell survival by IL-4 may not represent happenstance. Anergic B cells represent a cloaked version of the (less avid) autoreactive specificities excluded from the immunocompetent pool. Interference with the Fas deletional pathway may constitute a mechanism to reverse the fate of such B cells in situations

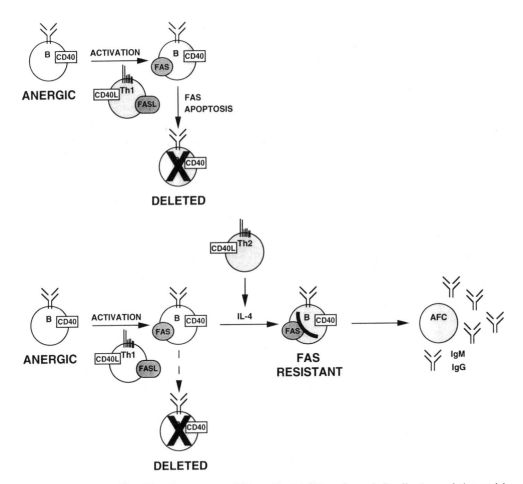

Figure 4. IL-4 secreted by Th2 cells may oppose Th1-mediated deletion of anergic B cells. A speculative model relating anergy, Fas-mediated apoptosis, Fas-resistance, and autoantibody formation is presented. In this model, anergic B cells are deleted (denoted by an "X") by FasL-expressing Th1 cells, possibly following activation by CD40L on the same effectors (top). This action may be opposed by IL-4 secreted by Th2 cells (that are relatively deficient in FasL expression) which induces a state of Fas-resistance in which Fas signaling for cell death is opposed as a result of intracellular changes (denoted by an arc)(bottom). In this way anergic B cells may escape the deletional pathway, perhaps to rejoin the immunocompetent pool, providing the opportunity for these autoreactive B cells to differentiate to antibody forming cells (AFC) that produce autoantibody. See text for additional details.

or at times where the autoreactive specificities are needed by the organism to counteract microbial pathogens whose antigenic determinants cross-react with self. At the risk of autoimmunity, IL-4-induced Fas-resistance may act as a fail-safe mechanism to cull from anergic B cells those specificities needed in extreme circumstances. This could represent another outcome of the cross-regulatory effects of Th1 and Th2 cells, inasmuch as Th1 cells express FasL but not IL-4, whereas Th2 cells express IL-4 but little FasL[45,46]. Alternatively, IL-4-induced Fas-resistance may represent an unintended side effect of the elevated levels of cytokines that may accompany situations of extreme immunological stress; the recognized increase in the frequency of autoantibodies with increasing age could be interpreted as reflecting the cumulative effects of many such episodes. A model for this is presented in Figure 4.

During immune responses antigen-specific B cells establish germinal centers. It has recently been reported that only the combination of sIg and CD40 engagement stimulates resting B cells to express the phenotype of germinal center cells[47]. As we and others have shown, this same combination of receptor triggering produces a stable state of Fas-resistance. It is known that germinal center cells express elevated levels of Fas but that positive selection of somatically mutating B cells does not depend on Fas-mediated apoptosis[48]. It may be speculated, then, that induction of Fas-resistance through antigen receptor triggering is an integral part of germinal center formation, insuring germinal center B cell viability against at least one form of death induction. As discussed here, sIg and IL-4R signals act in concert to establish strong Fas-resistance, and it may be that optimal B cell survival during germinal center formation and immune responses depends on simultaneous engagement of both receptors.

In sum, the regulation of Fas-mediated apoptosis by specific receptor signaling represents a new dimension to the regulation of immune responsiveness and autoreactive antibody formation, whose implications for normal- and auto-immunity are still being explored.

ACKNOWLEDGMENTS

This work was supported by United States Public Health Service grant AI40181 and by a grant from the Arthritis Foundation.

REFERENCES

1. S. Nagata, and P. Golstein. The Fas death factor. *Science* 267:1449. (1995).
2. Y. Nishimura, A. Ishii, Y. Kobayashi, Y. Yamasaki, and S. Yonehara. Expression and function of mouse Fas antigen on immature and mature T cells. *J. Immunol.* 154:4395 (1995).
3. R. Watanabe-Fukunaga, C.I. Brannan, N. Itoh, S. Yonehara, N.G. Copeland, N.A. Jenkins, and S. Nagata. The cDNA structure, expression and chromosomal assignment of the mouse Fas antigen. *J. Immunol.* 148:1274 (1992).
4. S. Nagata, and T. Suda. Fas and Fas ligand: *lpr* and *gld* mutations. *Immunol. Today* 16:39 (1995).
5. F. Rieux-Laucat, F. Le Deist, C. Hivroz, I.A.G. Roberts, K.M. Debatin, A. Fischer, and J.P. de Villartay. Mutations in Fas associated with human lymphoproliferative syndrome and autoimmunity. *Science* 268:1347 (1995).
6. G.H. fisher, F.J. Rosenberg, S.E. Straus, J.K. Kale, L.A. Middelton, A.Y. Lin, W. Strober, M.J. Lenardo, and J.M. Puck. Dominant interfering Fas gene mutations impair apoptosis in a human autoimmune lymphoproliferative syndrome. *Cell* 81:935 (1995).
7. L. B. Owen-Schaub, S. Yonehara, W. L. Crump III, and E. A. Grimm. DNA fragmentation and cell death is selectively triggered in activated human lymphocytes by Fas antigen engagement. Cell. Immunol. 140:197 (1992).

8. S.-T. Ju, H. Cui, D. J. Panka, R. Ettinger, and A. Marshak-Rothstein. Participation of target Fas protein in apoptosis pathway induced by CD4$^+$ Th1 and CD8$^+$ cytotoxic T cells. *Proc. Natl. Acad. Sci. USA* 91:4185 (1994).

9. Daniel, P. T., and P. H. Krammer. Activation induces sensitivity toward APO-1 (CD95)-mediated apoptosis in human B cells. *J. Immunol.* 152:5624. (1994).

10. Rothstein, T. L., J. K. M. Wang, D. J. Panka, L. C. Foote, Z. Wang, B. Stanger, H. Cui, S.-T. Ju, and A. Marshak-Rothstein. Protection against Fas-dependent Th1-mediated apoptosis by antigen receptor engagement in B cells. *Nature* 374:163 (1995).

11. C. Lagresle, P. Mondiere, C. Bella, P.H. Krammer, and T. Defrance. Concurrent engagement of CD40 and the antigen receptor protects naive and memory human B cells from APO-1/Fas-mediated apoptosis. *J. Exp. Med.* 183:1377 (1996).

12. L.C. Foote, T.J. Schneider, G.M. Fischer, J.K.M. Wang, B. Rasmussen, K.A. Campbell, D.H. Lynch. S.-T. Ju, A. Marshak-Rothstein, and T.L. Rothstein. Intracellular signaling for inducible antigen receptor-mediated Fas resistance in B cells. *J. Immunol.* 157:1878.

13. P. Lane, T. Brocker, S. Hubele, E. Padovan, A. Lanzavecchia, and F. McConnell. Soluble CD40 ligand can replace the normal T-cell derived CD40 ligand signal to B cells in T cell dependent activation. *J. Exp. Med.* 177:1209.

14. D.A. Francis, J.G. Karras, X. Ke, R. Sen, and T.L. Rothstein. Induction of the transcription factors NF-kB, AP-1 and NF-AT during B cell stimulation through the CD40 receptor. *Intl. Immunol.* 7:151 (1995).

15. Cifone, M.G., R. De Maria, P. Roncaioli, M.R. Rippo, M. Azuma, L.L. Lanier, A. Santoni, and R. Testi. Apoptotic signaling through CD95 (Fas/Apo-1) activates an acidic sphingomyelinase. *J. Exp. Med.* 177:1547 (1993).

16. E. Gulbins, R. Bissonnette, A. Mahboubi, S. Martin, W. Nishioka, T. Brunner, G. Baier, G. Baier-Bitterlich, C. Byrd, F. Lang, R. Kolesnick, A. Altman, and D. Green. FAS-induced apoptosis is mediated via a ceramide-initiated RAS signaling pathway. *Immunity* 2:341 (1995).

17. C. G. Tepper, S. Jayadev, B. Liu, A. Bielawska, R. Wolff, S. Yonehara, Y. A. Hannun, and M. F. Seldin. Role for ceramide as an endogenous mediator of Fas-induced cytotoxicity. *Proc. Natl. Acad. Sci. USA* 92:8443 (1995).

18. S. Cory. Regulation of lymphocyte survival by the Bcl-2 gene family. *Ann. Rev. Immunol.* 13:513 (1995).

19. Z. Wang, J.G. Karras, R.G. Howard, and T.L. Rothstein. Induction of bcl-x by CD40 engagement rescues sIg-induced apoptosis in murine B cells. *J. Immunol.* 155:3722 (1995).

20. P. Liang and A.B. Pardee. Differential display of eukaryotic messenger RNA by means of the polymerase chain reaction. *Science* 257:967 (1992).

21. J. Ohara, and W. E. Paul. B cell stimulatory factor BSF-1: Production of a monoclonal antibody and molecular characterization. *Nature* 315:333 (1985).

22. A. D. Keegan, K. Nelms, L.-M. Wang, J. H. Pierece, and W. E. Paul. Interleukin 4 receptor: signaling mechanisms. *Immunol. Today* 15:423 (1994).

23. J. C. Cambier, C. M. Pleiman, and M. R. Clark. Signal transduction by the B cell antigen receptor and its coreceptors. *Ann. Rev. Immunol.* 12:457 (1994).

24. L.B. Ivashkiv. Cytokines and STATs: How can signals achieve specificity? *Immunity* 3:1 (1995).

25. J.G. Karras, Z. Wang, S.J. Coniglio, D.A. Frank, and T.L. Rothstein. Antigen-receptor engagement in B cells induces nuclear expression of STAT5 and STAT6 proteins that bind and transacctivate an IFN-γ activation site. *J. Immunol.* 157:39.

26. J. J. Mond, N. Feuerstein, F. D. Finkelman, F. Huang, K.-P. Huang, and G. Dennis. B-lymphocyte activation mediated by anti-immunoglobulin antibody in the absence of protein kinase C. *Proc. Natl. Acad. Sci. USA* 84:8588 (1987).

27. K. Kawakami, and D. C. Parker. Antigen and helper T lymphocytes activate B lymphocytes by distinct signaling pathways. *Eur. J. Immunol.* 23:77 (1993).

28. M. P. Cooke, A. W. Heath, K. M. Shokat, Y. Zeng, F. D. Finkelman, P. s. Linsley, M. Howard, and C. C. Goodnow. Immunoglobulin signal transduction guides the specificity of B cell-T cell interactions and is blocked in tolerant self-reactive B cells. *J. Exp. Med.* 179:425 (1994).

29. J. M. Eris, A. Basten, R. Brink, K. Doherety, M. R. Kehry, and P. D. Hodgkin. Anergic self-reactive B cells present self antigen and respond normally to CD40-dependent T-cell signals but are defective in antigen-receptor-mediated functions. *Proc. Natl. Acad. Sci. USA* 92:4392 (1994).

30. C.C. Goodnow. Transgenic mice and analysis of B-cell tolerance. *Annu. Rev. Immunol.* 10:489 (1992).

31. S.E. Bell and C.C. Goodnow. A selective defect in IgM antigen receptor synthesis and transport causes loss of cell surface IgM expression on tolerant B lymphocytes. *EMBO J.* 1390:816 (1994).

32. D.A. Fulcher and A. Basten. Reduced life span of anergic self-reactive B cells in a double-transgenic model. *J. Exp. Med.* 179:125 (1994).

33. J. C. Rathmell, M. P. Cooke, W. Y. Ho, J. Grein, S. E. Townsend, M. M. Davis, and C. C. Goodnow. CD95 (Fas)-dependent elimination of self-reactive B cells upon interaction with CD4$^+$ T cells. *Nature* 376:181 (1995).

34. R.I. Tepper, D.A. Levinson, B.Z. Stanger, J. Campos-Torres, A.K. Abbas, and P. Leder. IL-4 induces allergic-like inflammatory disease and alters T cell development in transgenic mice. *Cell* 62:457 (1990).

35. E.M. Tan, P. Rodnan, and L. Garcia. Diversity of antinuclear antibodies in progressive systemic sclerosis. *Arthr. Rheum.* 23:617 (1978).

36. T.L. Rothstein. Signals and susceptibility to programmed cell death in B cells. *Curr. Opin. Immunol.* 8:362 (1996).

37. L. Zhang, R.G. Miller, and J. Zhang. Characterization of apoptosis-resistant antigen-specific T cells in vivo. *J. Exp. Med.* 183:2065 (1996).

38. J. Wang, I. Taniuchi, Y. Maekawa, M. Howard, M.D. Cooper, and T. Watanabe. Expression and function of Fas antigen on activated murine B cells. *Eur. J. Immunol.* 26:92 (1996).

39. J. Ogasawara, R. Watanabe-Fukunaga, M. Adachi, A. Matsuzawa, T. Kasugai, Y. Kitamura, N. Itoh, T. Suda, and S. Nagata. Lethal effect of the anti-Fas antibody in mice. *Nature* 364:806 (1993).

40. A. Fraser and G. Evan. A license to kill. *Cell* 85:781 (1996).

41. C.C. Goodnow. Balancing immunity and tolerance: Deleting and tuning lymphocyte repertoires. *Proc. Natl. Acad. Sci. USA* 93:2264 (1996).

42. C.C. Goodnow, R. Brink, and E. Adams. Breakdown of self-tolerance in anergic B lymphocytes. *Nature* 352:532 (1991).

43. C.C. Goodnow, J. Crosbie, H. Jorgensen, R.A. Brink, and A. Basten. Induction of self-tolerance in mature peripheral B lymphocytes. *Nature* 342:385 (1989).

44. B.A. Jacobson, D.J. panka, K.-A. Nguyen, J. Erikson, A.K. Abbas, and A. Marshak-Rothstein. Anatomy of autoantibody production: Dominant localization of antibody-producing cells to T cell zones in Fas-deficient mice. *Immunity* 3:509 (1995).

45. W.E. Paul and R.A. Seder. Lymphocyte responses and cytokines. *Cell* 76:241 (1994).

46. F. Ramsdell, M.S. Seaman, R.E. Miller, K.S. Picha, M.K. Kennedy, and D.H. Lynch. Differential ability of Th1 and Th2 T cells to express Fas ligand and to undergo activation-induced cell death. *Intl. Immunol.* 6:1545 (1994).

47. L. Galibert, N. Burdin, B. de Saint-Vis, P. Garrone, C. Van Kooten, J. Banchereau, and F. Rousset. CD40 and B cell antigen receptor dual triggering of resting B lymphocytes turns on a partial germinal center phenotype. *J. Exp. Med.* 183:77 (1996).

48. K.G.C. Smith, G.J. V. Nossal, and D.M. Tarlington. FAS is highly expressed in the germinal center but is not required for regulation of the B-cell response to antigen. *Proc. Natl. Acad. Sci. USA* 92:11628 (1995).

10

MURDER AND SUICIDE

A Tale of T and B Cell Apoptosis

David W. Scott,[*] Tommy Brunner, Dubravka Donjerković, Sergei Ezhevsky,
Terri Grdina, Douglas Green, Yufang Shi, and Xiao-rui Yao

Department of Immunology
Holland Laboratory for the Biomedical Sciences
American Red Cross
Rockville, Maryland 20855
La Jolla Institute for Allergy and Immunology
La Jolla, California 92037

1. ABSTRACT

When murine B-lymphoma cells are activated by crosslinking their membrane IgM receptors, they show evanescent myc transcription, growth arrest mediated by an increase in p27, and then undergo programmed cell death. Our laboratory has previously shown that the initiation of this process requires the activation of src-family protein tyrosine kinases (PTK) via their association with the ITAM (Immunoreceptor Tyrosine Activation Motif) in the Ig-associated proteins, Igα and Igβ. While PTK activation is required for growth arrest and apoptosis, it is not sufficient since mutation of critical tyrosines in the ITAM may allow for initial phosphorylation events, but no apoptosis. To explore the role of these kinases in apoptosis, we transfected both T cell and B cell lymphomas with CD8-PTK chimeric fusion proteins and found that only CD8-syk crosslinking led to cell death in the A1.1 T cell line; none of the B cell transfectants were responsive to CD8 crosslinking. Interestingly, many of the T cell transfectants showed strong PTK activation, but no measurable biologic response. Thus, initial signaling (chemistry) need not have cellular consequences (biology). To examine the role of c-myc in B cell apoptosis, we used antisense for c-myc and found that it prevented the loss of message for this oncogene and blocked apoptosis; antisense oligos for p27 also blocked cell cycle arrest and apoptosis. Interestingly, treatment of B cell lymphomas with dexamethasone augmented apoptosis induced by receptor crosslinking, whereas it blocked anti-TCR-mediated apoptosis in T cells. We propose that myc is critically regulated by dexamethasone in establishing the responsiveness of T versus B cell

[*] To whom correspondence should be addressed. This is publication #15 of the Holland Lab Immunology Department, American Red Cross and was supported by USPHS grants, CA55644 and AI29691 (DWS).

lines. In addition, receptor crosslinking led to FasL expression and suicide in T cell lines but not in B lymphomas. Although B cells are sensitive to Fas-mediated death, anti-IgM-driven B cell apoptosis appears to be Fas-independent. These data suggest that T and B cell apoptosis may be independently regulated to maintain the integrity of the immune system, and that myc transcription plays a pivotal role in this process.

2. INTRODUCTION AND OVERVIEW

The immune system must not only distinguish between foreign, potentially danger-ous entities and self antigens, it must also regulate the magnitude of both T and B cell re-sponses. It is well established that this is accomplished via programmed cell death/apoptosis. Thus, for example, studies with superantigens, in transgenic mice, and with both T and B cell lymphomas, provide evidence that deletional tolerance occurs by apoptosis initiated through specific TCR or Ig receptor ligation. Further data have substan-tiated that this process also occurs as a means of controlling extensive proliferation in the immune system. Hence, lymphocytes, like most other nucleated cells, must pass through a window of potential apoptotic cell death before progressing in the cell cycle and dividing. How this occurs in different lymphoid compartments and why it may be different in T and B lymphocytes are the topics for this chapter. Recent data suggest T cell apoptosis may in-volve a Fas (Apo-1, CD95):FasL interaction (Brunner et al., 1995; Ju et al., 1995; Dhein et al., 1995). Indeed, Fas and FasL are upregulated in T cell hybridomas upon T cell re-ceptor ligation and may function in an autocrine-loop to initiate the apoptosis. These data suggest that Fas on T cells may regulate repertoire selection and autoimmunity, but also the overall levels of T cells involved in an immune response. The role of Fas in B cell regulation is less clear (Ju et al., 1995; Scott, Grdina and Shi, 1996).

3. B CELL SIGNALING VIA THE Ig RECEPTOR COMPLEX

Neither B cell (Ig) nor T cell receptors (TCR) are able to transmit a biochemical sig-nal to the interior of the cell because these receptor molecules lack sizable intracellular do-mains. In the case of the B cell, its receptor complex (BCR) is composed of membrane immunoglobulin and, minimally, a disulfide-linked heterodimer consisting of the 34 kDa Igα (designated CD79a), and a 39 kDa Igβ subunit (called CD79b). This Igα/Igβ dimer is critical to directly couple the BCR complex to *src*-related protein tyrosine kinases (PTKs) and the Syk72 PTK known to be involved in the signal transmission (Hombach et al., 1988; Sakaguchi, et al., 1988; Yamanishi et al., 1991; Burkhardt et al., 1991; Campbell and Sefton, 1992; Hutchcroft et al., 1992; Lin and Justement, 1992). Within the cytoplasmic do-mains of Igα and Igβ are consensus sequences termed the Immunoreceptor Tyrosine-based Activation Motif (ITAM), a motif which is also found in TCR/CD3-ζ, γ, δ, ε; FcγRIII-γ; FcεRI-β and γ, for example (Reth, 1989; Alber et al., 1993; Flaswinkel and Reth, 1994).

The ITAM is characterized by a tandem duplication of a tyrosine-containing se-quence YxxL/I, whose role is as a target for phosphorylation and to "dock" other compo-nents of the complex for receptor activation (Flaswinkel and Reth, 1994; Pleiman et al., 1994). To examine the role of the ITAM and the kinases associated with this complex, we established a series of transfectants of chimeric fusion protein constructs for the study of inhibitory signaling pathways. This system took advantage of a series of B cell lym-phomas in which cross-linking of the BCR complex induces growth arrest and cell death by apoptosis (Scott et al., 1986; Pennell and Scott, 1986a; Pennell and Scott, 1986b) that is dependent on receptor-associated PTKs (Yao and Scott, 1993; Scheuermann et al., 1994), and downstream signaling pathways (Fischer et al., 1994; Joseph, et al., 1995; Ezhevsky et al., 1996). Using a series of cell lines transfected to express CD8 chimeras with Igα and Igβ cytoplasmic tails, we first showed that ligation with anti-CD8 antibody triggered signal cascades leading to growth arrest and apoptosis with both constructs (Yao et al., 1995), but not with a CD8:gamma chain cytoplasmic tail that lacked ITAMs. Furthermore, Y >> F mutations in the YxxL/I sequence of the Igα ITAM destroyed signaling for growth arrest and apoptosis by these chimeras despite evidence of phosphorylation in some cases (Table 1) (Yao et al., 1995). This suggests a critical role of the ITAM and that the phosphorylation process is necessary but not sufficient for apoptotic signaling.

Since we knew that the blk PTK was important in signal transduction in the CH31 lymphoma (Yao and Scott, 1993), we created a new series of chimeric constructs contain CD8 and either *src*-family kinases (blk, lyn, fyn, lck) or the syk or ZAP70 kinases. Surprisingly, although we found expression of all but the syk fusion proteins at high levels, crosslinking by anti-CD8 did not lead to growth arrest or apoptosis (Table 1). No viable CD8:syk clones were obtained. We propose that the lack of stable CD8:syk transfectants may be due to constitutive activation of this kinase in recipient cells leading to downstream signaling and cell death (see below). Transient assays and inducible promoters will be used to determine if this is the correct interpretation.

4. ITAMS, SRC KINASES, AND T CELL RECEPTOR SIGNALING: A SURPRISING CONTROL RESULT

As a control, all of these chimeric constructs were transfected into the A1.1 T cell hybridoma, which synthesizes IL-2 and undergoes apoptosis in the presence of anti-CD3 or

Table 1. Chimeric CD8 constructs used to examine signaling requirements for apoptosis

CD8:	Recipient cell	Bio-effect of anti-CD8	PY[*]
Igα	CH31	apoptosis	+
Igβ	CH31	apoptosis	+
Igα ITAM mutations	CH31		
M1 (Y->>F positions 23 and 34)		no effect	−
M2 (Y->>F positions 17[**] and 34)		no effect	+
M3 (Y->>F position 34)		no effect	+
M4 (Y->>F position 23)		no effect	+
M5 (Y->>F position 17[**])		apoptosis	+
M6 (deletion of 52 residues)		no effect	−
blk,lyn,fyn,lck,ZAP70	CH31	no effect	ND
syk	CH31[***]	—	
blk	A1.1	no effect	+
lyn	A1.1	no effect	−
fyn	A1.1	no effect	−
lck	A1.1	no effect	+
syk	A1.1	apoptosis	+
ZAP70	A1.1	no effect	−

[*]Initial phosphotyrosine activation as measured in anti-PY immunoblots.
[**]Residue outside the YXXL ITAM motif.
[***]No surviving clones obtained.

anti-TCR reagents (Shi et al., 1990; Scott et al., 1996). We obtained numerous clones of A1.1 expressing each of the fusion protein constructs, including CD8:syk. We, therefore, tested first whether we obtained similar receptor expression and then if anti-CD8 could drive IL-2 synthesis as well as anti-CD3. All of the T cell transfectants expressed similar levels of CD8 (and CD3) and these cell lines made IL-2 in response to anti-CD3 (Yao et al., in preparation, 1996). However, only the CD8:syk transfectants showed IL-2 production in the presence of anti-CD8. This was surprising since no IL-2 was obtained with CD8:ZAP70 or CD8:lck, both of which contain the cytoplasmic tails typical of T cell kinases. To test for signaling to cell death via apoptosis, we measured cell viability by the MTT assay and by phase microscopy, which showed that only the CD8:syk signaled for cell death (figure 1).

That this death was due to apoptosis was examined using propidium iodide to measure hypodiploid nuclei and by agarose gel electrophoresis of cellular DNA obtained after treatment with anti-CD8 or anti-CD3 for up to 24 hours. The results in figure 1**B** show typical patterns of internucleosomal DNA banding with anti-CD8, but only for the CD8:syk clones. All cells were equally susceptible to cell death via apoptosis with anti-CD3.

These results suggest that crosslinking of syk, but not ZAP70, in T cell lines can activate the signals for apoptosis, whereas none of the src-kinase chimeras were effective. To establish that crosslinking was effectively initiating the biochemical events necessary for apoptosis, we performed phosphotyrosine blotting with all of the cell lines transfected with these chimeric CD8 fusion proteins. Interestingly, crosslinking with anti-CD8 led to a strong phosphotyrosine (PY) signal is the CD8 chimeras with syk, as well as CD8:blk and CD8:lck (Yao et al., in preparation). Thus, one can observe a strong (bio)chemical signal without concomitant biological results, that is, growth arrest or cell death. No PY signal was observed with the ZAP70 chimera, either due to the need for more aggregation, lack of an associated phosphatase activity (such as provided by CD45), or the failure to complex with a needed src kinase, such as lck, for example. Additional doubly transduced cells lines are needed to test these possibilities. Minimally, however, our data suggest that CD8:syk crosslinking can activate the biochemical processes for apoptosis. A scheme to explain these findings in T cells and in B cells is presented in figure 2. Since syk is found in developing thymocytes, this system may provide a powerful model to examine the signals for apoptosis in the developing repertoire. Further studies of the proteins which associate with CD8:syk and differential substrate phosphorylation are underway.

5. SIGNALING VIA IgM VERSUS IgD: REGULATION OF CELL CYCLE PROGRESSION

We have previously reported that crosslinking of IgM but not IgD would lead to growth arrest and apoptosis in an IgD transfected derivative of the CH33 cell line (Ales-Martinez et al., 1988); Tisch and co-workers reported identical findings in the WEHI-231 lymphoma (Tisch et al., 1988). It is interesting to note that anti-IgM treatment leads to evanescent transcription of *c-myc*, with a drop to undetectable levels, followed by growth arrest and cell death in that order. Anti-IgD treatment also led to an increase in *c-myc* RNA above background steady state levels, but this message did not decrease to below background synthesis (Tisch et al., 1988; McCormack et al, 1984). We found that antisense for *c-myc* (directed at the first translation start site in exon 2) effectively protected B lymphomas from anti-IgM driven cell death, and that this treatment led to a stabilization of myc (Fischer et al., 1994). Similar results were reported by Sonenshein and colleagues (Mahesheran et al., 1992). Antisense for *c-myc* had no effect on the anti-IgD signal.

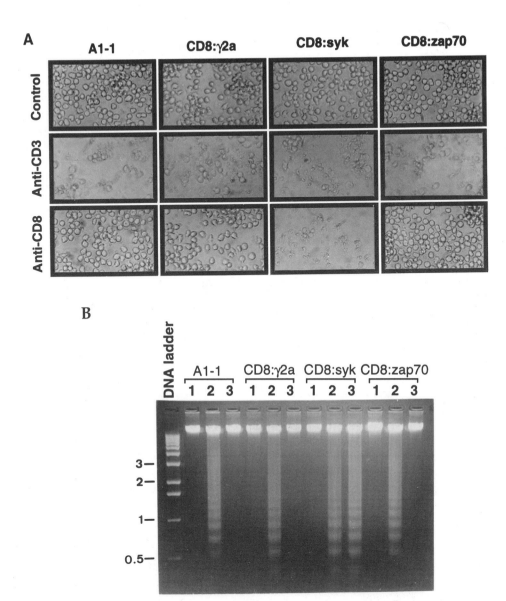

Figure 1. Induction of cell death in A1.1 T cell hybridoma cells expressing CD8 chimeric fusion proteins. Cells were transfected with constructs for fusion proteins consisting of the extracellular and transmembrane regions of CD8 with the cytoplasmic tails of γ2a heavy chain or blk, lyn, fyn, lck, syk or ZAP70. These were treated with anti-CD3 and anti-CD8 for 24 hours and analyzed for survivors by phase microscopy (A), as well as for apoptosis (B). **(A)** This figure shows that CD8:syk, but not the other chimeras, delivered a signal for cell death with anti-CD8. All clones were killed by anti-CD3, as expected. Note that only γ2a, syk and ZAP70 results are shown for simplicity, although none of the other clones was affected by anti-CD8. Similar results were observed with the MTT assay of cell viability. **(B)** Cell death is due to apoptosis in A1.1 T cell hybridoma cells expressing CD8 chimeric fusion proteins. Cells described above were analyzed for apoptosis by DNA gel electrophoresis. This figure shows that CD8:syk, but not the other chimeras, delivered a signal for apoptosis with anti-CD8. Similar results were observed using propidium iodide to assay for apoptosis.

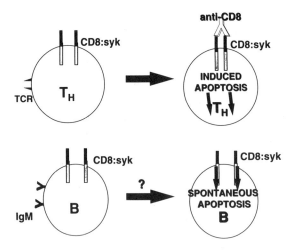

Figure 2. Scheme for induced and spontaneous cell death in A1.1 T cell hybridoma cells and CH31 B lymphoma cells expressing CD8:syk chimeric fusion. Cells were transfected with constructs for fusion protein consisting of the extracellular and transmembrane regions of CD8 with the cytoplasmic tail of syk, as in figure 1. These were treated with anti-CD8 for 24 hours (T cells) and analyzed for apoptosis. In the case of CH31 (B cells), spontaneous apoptosis was presumed to have occurred due to constitutive activation of CD8:syk leading to downstream consequences, i.e., cell death.

It should be noted that oligonucleotides that contain a CpG motif, which if flanked by a pyrimidine dimer are mitogenic for resting B cells (Krieg et al., 1995). We, therefore, wanted to establish whether mitogenicity of the oligos might account for the prevention of growth arrest and apoptosis, as seen with LPS (figure 3). This does not appear to be the case since nonsense or sense oligos containing this motif failed to reverse the effects of anti-IgM, and antisense oligos downstream from the translational start site (and lacking the mitogenic CpG motif) were effective (Fischer et al., 1995; Maddox and Scott, unpublished). Thus, the effects of antisense oligos for c-myc appear to be specific, although the reasons why they stabilize myc are not immediately apparent.

We know that anti-IgM crosslinking also causes cells to arrest in late G_1 (30), a result which appears to be due to a block in the serine/threonine phosphorylation of the *retinoblastoma* gene product, pRB (31). This, in turn, is due to the inhibition of pRB kinases, such as the complex formed by cdk2 and cyclin A (Joseph et al., 1995; Ezhevsky et al., 1996), as a result of the upregulation of the $p27^{kip1}$ kinase inhibitor (Ezhevsky et al., 1996). We recently examined whether anti-IgM and anti-IgD, which cause the same initial signals (PTK activation, phosphatidyl inositol breakdown, early c-myc upregulation) but have different biological consequences (death *versus* life) had the same effects on pRB via p27. We first observed (L. Joseph and D. W. Scott, unpublished) that anti-IgD treatment did not lead to a prevention of the phosphorylation of pRB, as anti-IgM does. To determine if this effect was due to differential effects on p27 synthesis, we performed western blots for p27 in extracts of IgM[+], IgD[+] lymphomas treated with either of these anti-Ig reagents. Our results showed (Donjerković and Scott, unpublished and figure 4**A**) that anti-IgM consistently caused an upregulation of p27, whereas anti-IgD treatment did not. These data suggest that regulation of p27 synthesis may be critical in the control of B lymphoma growth and apoptosis. To formally test this hypothesis, we treated B lymphoma cells with antisense oligos for p27 before treatment with anti-IgM and found (figure 4**B**) a complete blockade of growth arrest and apoptosis. However, since this oligonucleotide contained a consensus mitogenic motif, it was possible that the effects were secondary to the antisense design. We then tested alternative oligos and found that the only effective ones contained the mitogenic motif. Therefore, this effect may not be due to prevention of p27 synthesis. Experiments are in progress with stable transfectants to further examine the role of p27 in cell cycle blockade and apoptosis.

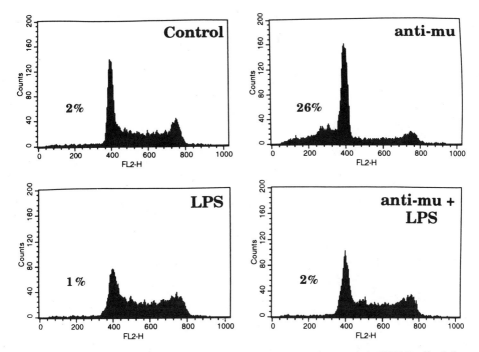

Figure 3. Effect of mitogenic LPS on anti-IgM mediated growth arrest and apoptosis in CH31 B cells. Cells were treated with anti-IgM for 20 hours in the presence or absence of LPS. Apoptosis was assayed using propidium iodide and flow cytometry.

6. ROLE OF Fas:FasL INTERACTIONS IN B LYMPHOMA APOPTOSIS

Recent data suggest that T cells can be driven to commit "suicide" by upregulating the expression of both Fas (CD95) and its ligand (FasL). Thus, under appropriate conditions of activation (for example, with anti-CD3), T cells synthesize and release soluble FasL, which can then bind to and signal via membrane Fas, leading to apoptosis (Brunner et al., 1995; Ju et al., 1995; Dhein et al., 1995). To test whether B cells use the same pathway, we and others (Onel et al., 1995) examined the expression of Fas on both resting and anti-IgM activated B cells. CD95 expression was low on resting B cells, although it was increased upon strong mitogenic stimulation (Onel et al., 1995). In WEHI-231 lymphoma cells, Fas antigen expression is also low, but does not increase with anti-IgM crosslinking (Onel et al., 1995; Rothstein et al., 1995). Importantly, crosslinking with anti-IgM led to no detectable increase in Fas ligand mRNA or protein (4 and T. Brunner, D. Green, T. Grdina and D.W. Scott, unpublished). In addition, incubation of WEHI-231 or CH31 cells with a soluble Fas-Fc fusion protein failed to protect these cells from activation-induced apoptosis with anti-IgM (Scott, D. et al. , 1996; Rothstein et al., 1995), whereas this form of Fas-Fc blocked anti-CD3-driven cell death in a T cell hybridoma (Brunner et al., 1995; Scott, D. et al. , 1996; Rothstein et al., 1995 and figure 5).

CH31 and WEHI-231 are also TNFα-insensitive (X-R. Yao, unpublished); therefore, we conclude that B cell lymphomas do not invoke an autocrine Fas or TNF pathway to commit suicide, as T cell appear to do. Thus, B cell apoptosis, with the caveat that we have only tested these anti-IgM-sensitive lymphomas, is Fas-independent and different

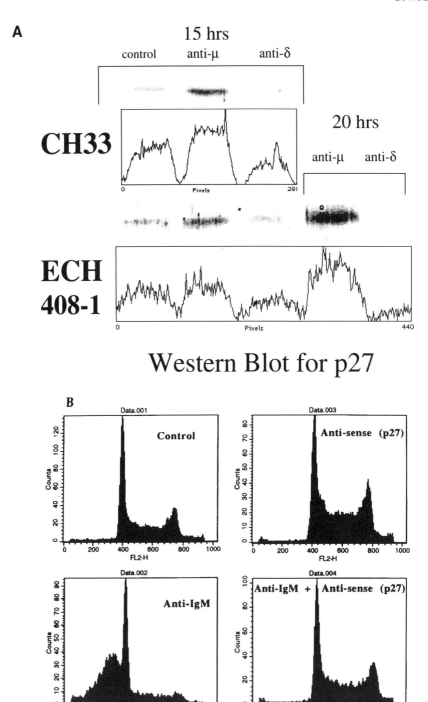

Figure 4. Role of the p27 kinase inhibitor in B lymphoma growth arrest and apoptosis. (**A**) Densitometry tracing of (western) immunoblot of ECH408 and CH33 B lymphoma cells treated with anti-IgM or anti-IgD for up to 20 hours. These results show that anti-IgM causes an upregulation of p27, whereas anti-IgD treatment does not, consistent with their effects on growth arrest. (**B**). Effect of antisense oligos to p27 on anti-IgM mediated apoptosis. CH31 cells were treated for 20 hours with anti-IgM plus or minus antisense oligonucleotides for p27. Apoptosis was assayed using propidium iodide and flow cytometry. These data show that p27 may be critical in the control of B lymphoma growth and apoptosis; however, the effect may be indirect due to mitogenic motifs in the antisense oligonucleotides.

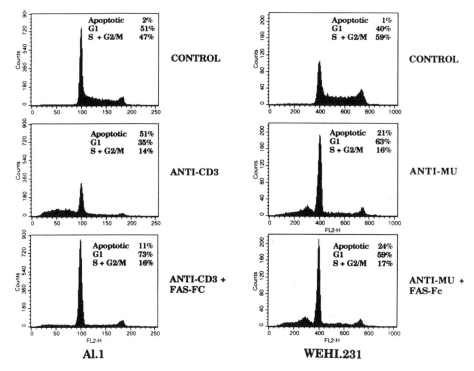

Figure 5. Role of the Fas and FasL in receptor-mediated apoptosis in B and T cells. A1.1 T cells (left side) and WEHI-231 B cells (right side) were cultured overnight with anti-receptor antibodies (anti-CD3 or anti-IgM, respectively) plus or minus soluble Fas-Fc fusion protein and apoptosis assayed using propidium iodide and flow cytometry. Fas-Fc blocked T cell apoptosis but did not interfere with B cell activation-induced cell death.

from the suicide pathway in T cells. It is important to note that recent data from several labs indicate that Fas is expressed at multiple B-lymphocyte stages of development but that apoptosis in germinal centers may be Fas-independent (Garrone et al., 1995; Mandik et al., 1995; Lagresle, et al., 1995; Smith et al., 1995; Hann et al., 1995). Thus, it is possible that B cells can be murdered by activated T cells under the correct circumstances, although they do not appear to "commit suicide" by this pathway (see Scott, D. et al., 1996).

7. THE DILEMMA OF CD40 TRIGGERING!

CD40 interaction with gp39 (CD40L) on T cells, in the presence of the appropriate cytokines causes B cells to enter the cell cycle, differentiate and switch to downstream Ig isotypes. Moreover, in models of B cell apoptosis, CD40 ligation (either with anti-CD40 or membranous CD40L) will protect these cells from death signals (Valentine and Licciardi, 1992; Heath et al., 1994; Cleary et al., 1995; Scott et al., 1995). Ironically, CD40 stimulation also leads to an upregulation of Fas expression on B cells, thus rendering them more sensitive to T cell-derived FasL-mediated cell death. As noted above, this can lead to B cell murder! How do we account for this sensitivity to Fas-mediated death and still achieve optimal immune responsiveness? Are any of these processes explicable by invoking the Fas pathway described above?

As stated above, we have found that B cells may express Fas, but they do not upregulate FasL in response to anti-IgM signaling (Brunner, T., Grdina, T., Green, D. and

Scott, D., unpublished). Thus, they do not appear to die via a Fas:FasL autocrine pathway. Nonetheless, B cells can be killed (murdered) by T cells expressing FasL (Rathmell et al., 1995)! Recent data from T. Rothstein's lab suggests that anti-IgM signaling may in fact protect B cells from FasL-mediated death (see Rothstein, this volume and Rothstein et al., 1995). Moreover, CD40 crosslinking also prevents the upregulation of p27 in murine B lymphomas (T. Tsubata, personal communication, 1996), a result we recently confirmed, and we know that p27 upregulation appears to be a prerequisite for growth arrest and apoptosis in this model. The modulation of FasL-mediated death signals upon BCR ligation, coupled with the CD40-driven protection from apoptosis, provides novel control mechanisms for B cell responsiveness which may also be dependent on the kinetics and milieu in which B:T interactions occur *in vivo* (*e.g.*, in germinal centers), and whether T_H1 or T_H2 cells are present.

Further studies in our lab have shown that apoptosis in B and T cells differs not only in terms of the stimuli and involvement of Fas, but also at the level of their oncogene signaling pathways. These are summarized in the next section.

8. SYNERGY AND ANTAGONISM IN RESPONSE TO CORTICOSTEROIDS AND ANTIGEN RECEPTOR CROSSLINKING

We, therefore, asked the following question: Are T cells and B cells fundamentally different in the pathways they use to commit suicide? Clearly, there is a difference in the expression of Fas or FasL and their upregulation or utilization in suicide. What other pathways are involved in apoptosis? For example, in T cell lymphomas or immature thymus cells, glucocorticoids, such as dexamethasone, or antibodies to the TCR complex can induce apoptosis. However, when these two signals are provided together to the T cell, they antagonize each other and block apoptosis (Zacharchuk et al., 1990). Glucocorticoids alone induced minimal growth arrest and apoptosis in the CH31 or WEHI-231 B cell lymphomas. In contrast to the observations with T cells, IgM receptor crosslinking and glucocorticoids synergized rather than antagonized cell death signals (Scott et al., 1996). Earlier, we found that TGF-β could synergize with anti-IgM to increase apoptosis in B lymphoma cells (Warner et al., 1992), while it is an inhibitor of activation-induced apoptosis in T cell hybridomas (Y. Shi, personal communication).

Coupled with the Fas-independence of B cell suicide, these data suggest fundamental differences in the regulation of apoptosis in B and T lymphocytes. We propose that the basis of these differences resides in the regulation of the *c-myc* oncogene. In T cells, antisense oligonucleotides against *c-myc* block apoptosis by preventing the upregulation of *c-myc* transcription, while in B cells, we found that antisense myc led to stabilization of both myc message and protein (Fischer et al., 1994). Recent data from Sonenshein's group (Wu, et al., 1996) validate that loss of myc is more critical to the survival of B cell lymphomas. This clearly contrasts with the situation in T cell hybridomas and fibroblasts, in which overexpression of c-myc leads to programmed cell death (Shi et al., 1992). However, it should be noted that treatment of B cell lymphomas with TGF-β can lead to apoptosis, which is a consequence of the loss of myc (Fischer et al., 1994). In fact, TGF-β does not cause even an evanescent increase in c-myc transcription in B cell lymphomas. Akin to the situation with corticosteroids, anti-IgM plus TGF-β can synergize to cause rapid apoptosis in these B cell lymphomas (Warner and Scott, 1992). Inhibitors of transcription and translation, such as actinomycin D and cycloheximide, also induce apoptosis in B cell lymphomas but block programmed cell death in T cells. Therefore, we propose that apoptotic cell death in B cells is tightly regulated by short-lived inhibitory proteins that are either myc-dependent or regulated by myc. This is testable.

In conclusion, it is clear that both B cells and T cells receive signals for apoptosis in both deletional tolerance and the regulation of immune responsiveness. These occur via receptor-associated protein tyrosine kinases that lead to differing downstream effector mechanisms. In T cells, FasL synthesis leads to cell death in an autocrine fashion. While sensitive to Fas-mediated signals, B cell death appears to be a consequence of growth arrest and c-myc dysregulation. The interdependence of B and T cells in the immune system may require that these lymphocyte populations regulate themselves differentially to maintain the integrity of this system and the organism.

REFERENCES

Alber, G., Kim, K.-M., Weiser, P., Riesterer, C., Carsetti, R., and Reth, M. 1993 Molecular mimicry of the antigen receptor signaling motif by transmembrane proteins of Epstein-Barr virus and the bovine leukemia virus. Curr. Biol. 3, 333–339.

Ales-Martínez, J-E., Warner, G. and Scott, D.W. 1988. Immunoglobulin D and M mediate signals that are qualitatively different in B cells with an immature phenotype. Proc. Natl. Acad. Sci. USA 85, 6919–6923.

Brunner, T., Mogil, R.J., LaFace, D., Yoo, N.J., Mahboubi, A., Echeverri, F., Martin, J., Force, W.R., Lynch, D.H., Ware, C.F., and Green, D.R. 1995. Cell-autonomous Fas (CD95)/Fas-Ligand and interaction mediates activation-induced apoptosis in T cell hybridomas. Nature 373, 441–444.

Burkhardt, A.L., Brunswick, M., Bolen, J.B., and Mond, J.J. 1991. Anti-immunoglobulin stimulation of B lymphocytes activate src-related protein-tyrosine kinases. Proc. Natl. Acad. Sci. USA 88, 7410–7414.

Campbell, M.-A., and Sefton, B.M. 1992. Association between B-lymphocyte membrane immunoglobulin and multiple members of the src family of protein tyrosine kinases. Mol. Cell. Biol. 12, 2315–2321.

Cleary, A.M., Fortune, S.M., Yellin, M.J., Chess, L., and Lederman, S. 1995. Opposing roles of CD95 (Fas/APO-1) and CD40 in the death and rescue of human low density tonsillar B cells. J. Immunol. 155, 3329–3337.

Dhein, J., Walczak, H., Baümler, C., Debatin, K.-M., and Krammer, P. 1995. Autocrine T-cell suicide mediated by APO-1 (Fas/CD95). Nature 373, 438–441.

Ezhevsky, S.A., Toyoshima, H., Hunter, T., and Scott, D.W. 1996. Role of cyclin A and p27 in anti-IgM-induced G1 growth arrest of murine B-cell lymphomas. Mol. Biol. Cell 7, 553–564.

Fisher, G., Kent, S.C., Joseph, L., Green, D.R., and Scott, D.W. 1994. Lymphoma models for B cell activation and tolerance. X. Anti-μ-mediated growth arrest and apoptosis of murine B cell lymphomas is prevented by the stabilization of myc. J. Exp. Med. 179, 221–228.

Flaswinkel, H., and Reth, M. 1994. Dual role of the tyrosine activation motif of the Ig-α protein during signal transduction via the B cell antigen receptor. EMBO J. 13, 83–89.

Garrone, P., Neidhardt, E.-M., Garcia, E., Galibert, L., van Kooten, C., and Banchereau, J. 1995. Fas ligation induces apoptosis of CD-40-activated human B lymphocytes. J. Exp. Med. 182, 1265–1273

Han, S., Zheng, B., Dal Porto, J., and Kelsoe, G. 1995. In situ studies of the primary immune response to (4-hydroxy-3-nitrophenyl) acetyl. IV. Affinity-dependent, antigen-driven B cell apoptosis in germinal centers as a mechanism for maintaining self-tolerance. J. Exp. Med. 182, 1635–1644.

Heath, A.W., Wu, W.W., and Howard, M.C. 1994. Monoclonal antibodies to murine CD40 define two distinct functional epitopes. Eur. J. Immun. 24, 1828–1834.

Hombach, J., Leclercq, L., Radbruch, A., Rajewsky, K., and Reth, M. 1988. A novel 34-kDa protein co-isolated with the IgM molecule in surface IgM expressing cells. EMBO J. 7, 3451–3456.

Hutchcroft, J.E., Harrison, M.L., and Geahlen, R.L. 1992. Association of the 72-kDa protein tyrosine kinase PTK72 with the B cell antigen receptor. J. Biol. Chem. 267, 8613–8619.

Joseph, L.F., Ezhevsky, S., and Scott, D.W. 1995. Lymphoma model for B-cell activation and tolerance: Anti-immunoglobulin M treatment induces growth arrest by preventing the formation of an active kinase complex which phosphorylates retinoblastoma gene product in G_1, Cell Growth & Differentiation 6, 51–57.

Ju, S.-T., Panka, D., Cui, H., Ettinger, R., el-Khatib, M., Sherr, D., Stanger, B.Z., and Marshak-Rothstein, A. 1995. Fas (CD95)/FasL interactions required for programmed cell death after T-cell activation. Nature 373, 444–448.

Krieg, A.M., Yi, A-K., Matson, S., Waldschmidt, T.J., Bishop, G.A., Teasdale, R., Koretzky, G.A., and Klinman, D.M. 1995. CpG motifs in bacterial DNA trigger direct B-cell activation. Nature 374, 546–549.

Lagresle, C., Bella, C., Daniel, P.T., Krammer, P., and DeFrance, T. 1995. Regulation of germinal center B cell differentiation. Role of the human APO-1/Fas (CD95) molecule. J. Immunol. 154, 5746–5756.

Lin, J., and Justement, L.B. 1992. The mb-1/b29 heterodimer couples the B cell antigen receptor to multiple *src* family protein tyrosine kinases. J. Immunol. 149, 1548–1555.

Maheswaran, S., McCormack, J.E., and Sonenshein, G.E. 1993. Changes in phosphorylation of myc oncogene and RB antioncogene protein products during growth arrest of the murine lymphoma WEHI 231 cell line. Oncogene 6, 1965–1971.

Mandik, L., Nguyen, K.-A.T., and Erickson, J. 1995. Fas expression on B lineage cells. Eur. J. Immunol. 25, 3148–3154.

McCormack, J.E., Pepe, V.H., Kent, R.B., Dean, M., Marshak-Rothstein, A., and Sonenshein, G.E. 1984. Specific regulation of c-myc oncogene expression in a murine B-cell lymphoma. Proc. Natl. Acad. Sci, USA 81, 5546–5550.

Onel, K.B., Tucek-Szabo, C.L., Ashany, D., Lacy, E., Nikolic-Zugic, J., and Elkon, K.B. 1995. Expression and function of the murine CD95/FasR/APO-1 receptor in relation to B cell ontogeny. Eur. J. Immunol. 25, 2940–2947.

Pennell, C.A., and Scott, D.W. 1986a. Models and mechanisms for signal transduction in B cells. Immunol. Res. 5, 61–70.

Pennell, C.A., and Scott, D.W. 1986b. Lymphoma models for B cell activation and tolerance. IV. Growth inhibition by anti-Ig of CH31 and CH33 B-lymphoma cells. Eur. J. Immunol. 16, 1577–1581.

Pleiman, C.M., Abrams, C., Gauen, L.T., Bedzyk, W., Jongstra, J., Shaw, A.S., and Cambier, J.C. 1994. Distinct p53/p56^{lyn} and p59^{lyn} domains associate with non-phosphorylated and phosphorylated Ig-α. Proc. Natl. Acad. Sci. USA 91, 4268–4272.

Rathmell, J.C., Cooke, M.P., Ho, W., Grein, J., Townsend, S., Davis, M.M., and Goodnow, C. 1995. CD95 (fas)-dependent elimination of self-reactive B cells upon interaction with CD4^{+} T cells. Nature 376, 181–183.

Reth, M. 1989. Antigen receptor tail clue. Nature 338, 383–384.

Rothstein, T.L., Wang, J. K. , Panka, D.J., Foote, L.C., Wang, Z., Stanger, B., Cui, H., Ju, S.T., and Marshak-Rothstein, A. 1995. Protection against Fas-dependent TH1-mediated apoptosis by antigen receptor engagement in B cells. Nature 374, 163–165.

Sakaguchi, N., Kashiwamura, S., Kimoto, M., Thalmann, P., and Melchers, F. 1988. B lymphocyte lineage-restricted expression of *mb-1*, a gene with CD3-like structural properties. EMBO J. 7, 3457–3464.

Scheuermann, R.H., Racila, E., Tucker, T., Yefenof, E., Street, N.E., Vitetta, E. S., Picker, L.J., and Uhr, J.W. 1994. Lyn tyrosine kinase signals cell cycle arrest but not apoptosis in B-lineage lymphoma cells. Proc. Natl. Acad. Sci. USA 91, 4048–4052.

Scott, D.W., Livnat, D., Pennell, C.A., and Keng, P. 1986. Lymphoma models for B cell activation and tolerance. III. Cell cycle dependence for negative signaling of WEHI-231 B lymphoma cells by anti-μ. J. Exp. Med. 164, 156–164.

Scott, D.W., Ezhevsky, S., Maddox, B., Washart, K., Yao, X., and Shi, Y. 1995. Scenes from a short life: Checkpoints and progression signals for immature B-cell life versus apoptosis. In: *Lymphocyte Signalling* (ed. K. Rigley and M. Harnett), John Wiley, Chichester, England, pp. 167–181.

Scott, D., Grdina, T., and Y. Shi. 1996. T cells commit suicide, but B cells are murdered. J. Immunol. 156, 2352–2356.

Shi, Y., Glynn, M., Guilbert, L.J., Cotter, T.G., Bissonette, R.P., and Green, D.R. 1992. Role for c-myc in activation induced apoptotic cell death in T-cell hybridoma. Science 257, 212–214.

Smith, K.G.C., Nossal, G.J.V., and Tarlinton, D.M. 1995. Fas is highly expressed in the germinal center but is not required for regulation of the B-cell response to antigen. Proc. Nat'l. Acad. Sci., USA 92, 11628–11632.

Tisch, R., Roifman, C.M., and Hozumi, N. 1988. Functional differences between immunoglobulins M and D are expressed on the surface of an immature B-cell line. Proc. Natl. Acad. Sci, USA 85, 6914–6918.

Valentine, M.A., and Licciardi, K.A. 1992. Rescue from anti-IgM-induced programmed cell death by the B cell surface proteins CD20 and CD40. Eur. J. Immunol. 22, 3141–3148.

Warner, G.L., Ludlow, J., Nelson, D.O., Gaur, A., and Scott, D.W. 1992 Anti-immunoglobulin treatment of B cell lymphomas induce active TGF-β but pRB hypophosphorylation is TGF-β independent. Cell Growth Differ. 3, 175–181.

Warner, G.L., Nelson, .D., and Scott, D.W. 1992. Synergy between TGF-beta and anti-IgM in growth inhibition of CD5^{+} B cell lymphomas. Ann. N. Y. Acad. Sci. 651, 274–276.

Wu, M., Arsura, M., Bella, R., Fitzgerald, M., Lee, H., Schauer, S., Sherr, D., and Sonenshein, G. 1996. Inhibition of c-myc expression induces apoptosis of WEHI 231 murine B cells. Molec. Cell. Biol. 16, 5015–5025.

Yamanashi, Y., Kakiuchi,T., Mizuguchi, J., Yamamoto, T., and Toyoshima, K. 1991. Association of B cell antigen receptor with protein tyrosine kinase Lyn. Science 251, 192–194.

Yao, X., and Scott, D.W. 1993a. Antisense oligodeoxynucleotides to *blk* gene prevents anti-μ-induced growth inhibition and apoptosis in a B-cell lymphoma. Proc. Natl. Acad. Sci. USA 90, 7964–7968.

Yao, X., and Scott, D.W. 1993b. Expression of protein tyrosine kinases in the Ig complex of anti-μ-sensitive and anti-μ-resistant B-cell lymphomas: Role of the p55^{blk} kinase in signaling growth arrest and apoptosis. Immunol. Rev. 132, 163–186.

Yao, X., Flaswinkel, H., Reth, M., and Scott, D.W. 1995. Immunoreceptor tyrosine -based activation motif is required to signal pathways of receptor-mediated growth arrest and apoptosis in murine B lymphoma cells. J. Immunol. 155, 652–661.

Zacharchuk, C.M., Mercep, M., Chakraborti, P.K., Simons, S.S., and Ashwell, J.D. 1990. Programmed T lymphocyte death. Cell activation- and steroid-induced pathways are mutually antagonistic. J. Immunol. 145, 4037–4045.

11

MOLECULAR MECHANISMS OF T LYMPHOCYTE APOPTOSIS MEDIATED BY CD3

Dexian Zheng,[1] Yanxin Liu,[1] Yong Zheng,[1] Ying Liu,[1] Shilian Liu,[1] Baoping Wang,[2] Markus Metzger,[2] Emiko Mizoguchi,[2] and Cox Terhorst[2]

[1]Chinese Academy of Medical Sciences
Peking Union Medical College
Beijing 100005, China
[2]Harvard Medical School
Boston, Massachusetts 02115

1. INTRODUCTION

Apoptosis or programmed cell death has been recently known as an very important biological phenomenon in development, differentiation and cellular homeostasis and constant size of organs of multicellular organisms. Lymphoid system is one of the best examples to demonstrate that apoptosis exists all the way through the lives of T and B cells. The T cell repertoire consists of an extremely diverse population of surface receptors (TCRα/β or TCRγ/δ) which participate in highly specific responses to different antigens and mediate antigen stimulation signals. Variable TCRα/β and TCRγ/δ receptors are associated with invariable CD3γ, δ, ϵ and ζ proteins, thus forming the TCR/CD3 complex.

CD3 peptides are expressed in all TCR$^+$ T cells and also found in fetal TCR$^-$ natural killer (NK) cells, but CD3ϵ, ζ polypeptides are only found in mature NK cells[1]. CD3γ, δ are glycoproteins, while CD3ϵ, ζ are simple protein molecules. The CD3ζ chain has a relatively short extracellular domain of only 9 amino acids and a long cytoplasmic tail of 112 amino acids[2].

It has been known that the activation signals of CD3ϵ and CD3ζ are transduced by immuno-receptor tyrosine-based activation motifs (ITAM) in their cytoplasmic tails. The tyrosine residues in the ITAM of both CD3ζ and CD3ϵ are phosphorylated during the activation process[3]. Stimulation of the T cells via the cytoplasmic tails of either CD3ϵ or CD3ζ resulted in tyrosine phosphorylation with quantitatively distinct patterns, suggesting that individual ITAM might play unique roles in TCR responses. Anti-CD3ϵ monoclonal antibody can trigger cell death by apoptosis in immature thymocytes and activation/proliferation responses or activation-induced cell death in mature and transformed T cells[4]. Recently, Combadiere et al.[5] reported that the CD3ζ was especially effective at promoting mature T cell apoptosis compared with the CD3ε, δ or γ chains. Apoptosis induced by single chain chimeras revealed that the first CD3ζ ITAM induced obviously more apoptosis than the third CD3ζ ITAM, while the second ζ ITAM was not involved in triggering apoptosis.

The evidences reported in the literature revealed that there might be two distinct signaling pathways: one mediated by CD3ζ and the other mediated by CD3ε chains. However, the role of the single chain of CD3ε in apoptosis induction and molecular mechanisms of the apoptotic signaling pathway are poorly understood. In this presentation, a preliminary result in this regard by using the transgenic mouse models and a single chain chimera by fusing the extracellular and transmembrane domains of human CD8α to the intracellular domain of human CD3ε will be reported.

2. THE ROLE OF CD3ε MOLECULE IN THE T LYMPHOCYTE APOPTOSIS

Wang et al.[6] has shown that a severe immunodeficiency with total loss of T and NK cells and a very small thymus (10–15% size of the wild type mice) was observed in the transgenic mice which contain more than thirty copies of the human CD3ε gene, indicating that CD3ε plays an important role in regulation of the development of the above two cell populations. The question we asked was whether the aberrant intracellular signal transduction through CD3ε cytoplasmic tail in thymocytes resulting in T and NK cell apoptosis was the cause of this pathological phenomenon. Histological studies of thymus of the fetal transgenic mice by using in situ TdT-mediated dUTP nick end labeling (TUNEL) showed that there was no difference of spontaneous apoptosis occurred between the CD3ε$^{+/+}$ fetal transgenic mice and the wild type litter mates except that the thymus was much smaller in the transgenic mice (10–30% size of the wild type), suggesting that CD3ε overexpression abrogated the thymus development in the transgenic mice. However the spontaneous apoptosis increased rapidly with age in the young transgenic mice as compared with that of the wild type, similar phenomenon was found in the CD3ζ$^{-/-}$ mice (Table. 1), indicating that both abundance of CD3ε and absence of CD3ζ molecules in thymocytes of the young transgenic mice facilitated spontaneous apoptosis.

Table 1. Spontaneous thymocyte apoptosis of the transgenic mice

Mice	Age (days)	Apoptotic thymocytes %
wt	2	0.334 ± 0.334
	28	0.930 ± 0.249
	90	1.356 ± 0.433
CD3ε$^{+/+}$	2	1.480 ± 0.697
	28	1.650 ± 0.786
	90	2.464 ± 0.469
CD3ζ$^{-/-}$	2	N.D.
	28	2.330 ± 0.644
	90	4.806 ± 0.370

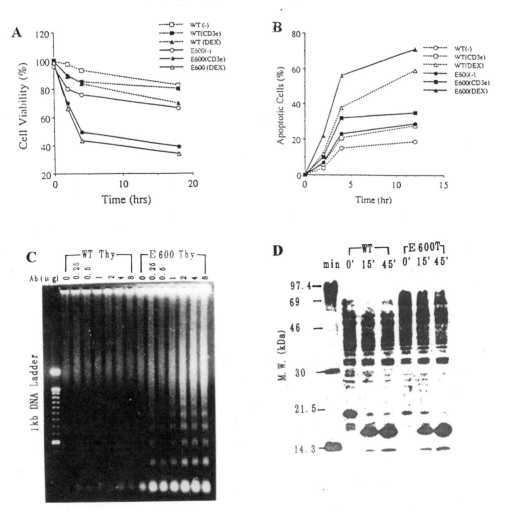

Figure 1. Activation-induced cell death of the thymocytes from the CD3ε[+/+] transgenic mice. A: Viability of thymocytes. The cells were treated with 2C11 or DEX and stained with Trypan blue; B: Apoptosis of the thymocytes. The cells were treated with 2C11 or DEX and stained with Hoechst and PI, analyzed by flow cytometry; C: DNA ladder. Cells were treated with 2C11 at differnt concentration. The DNA was extracted as routine procedure and run on 1% agarose gel; D: Western blot analysis. The thymocytes were treated with 2C11 for different time and the lysates were run on SDS-PAGE. The proteins were blot on nitrocellulose filter and probed with anti-phospho-tyrosine antibody. WT: the wild type mice. E600: CD3ε[+/+] mice. Thy: thymocytes.

Further studies were carried out to analyze the differences of activation-induced cell death (AICD) in thymocytes between the transgenic mice and the wild type mice. The thymocytes were isolated from the transgenic mice (E600) at the age of six weeks and treated with anti-CD3ε monoclonal antibody (2C11, IgG1, at final concentration of 5μg/ml) and dexamethesone (DEX, 10^{-6} M) followed by staining with Trypan blue. As shown in the Fig. 1A that thymocyte viability of the transgenic mice was lower than that of the wild type controls treated with either 2C11 or DEX. Hoechst staining and propidium iodine (PI) exclusion showed that cells died were by apoptosis and the apoptotic cell numbers of thymocytes from the transgenic mice were more than those of the

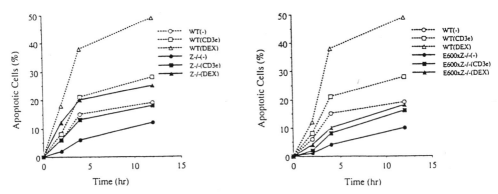

Figure 2. Activation-induced cell death of the thymocytes from the CD3$\zeta^{-/-}$ mice and the mice of CD3$\zeta^{-/-}$ crossed with CD3$\varepsilon^{+/+}$ mice. The cells were treated with 2C11 or DEX and stained with Hoechst and PI , analyzed by flow cytometry. WT: the wild type mice. E600: CD3$\varepsilon^{+/+}$ mice. Z$^{-/-}$: CD3ζ knockout mice (CD3$\zeta^{-/-}$). E600 x Z$^{-/-}$: the mice of CD3$\varepsilon^{+/+}$ crossed with CD3$\zeta^{-/-}$.

wild type (Fig. 1B), suggesting that the thymocytes of the transgenic mice were more sensitive to apoptosis induced by either 2C11 or DEX. Moreover, since DEX is a nonspecific apoptosis-inducer and the thymocytes of transgenic mice were also sensitive to DEX than those of the wild type, it is reasonable to predict that there might be an overlapping in the 2C11 and DEX mediated apoptosis pathways in the transgenic mice. The DNA fragmentation analysis showed that DNA ladder, the typical biochemical characteristic of apoptosis, occurred earlier in the CD3$\varepsilon^{+/+}$ mice than that seen in mice of the wild type (Fig. 1C). Western blot analysis revealed that phosphorylation of a protein was decreased in thymocytes of the transgenic mice treated with anti-CD3ε antibody, suggesting that sensitivity to apoptosis of the thymocytes in CD3$\varepsilon^{+/+}$ mice might be related to dephosphorylation of this unknown protein (Fig. 1D). The nature of this protein remains to be characterized.

In contrast to the CD3$\varepsilon^{+/+}$ transgenic mice, it was very interesting to note that thymocytes of the CD3$\zeta^{-/-}$ mice were resistant to apoptosis induced by anti-CD3ε or DEX (Fig. 2A). Levelt et al. demonstrated (personal communication) that administration of anti-CD3ε antibody to CD3$\zeta^{-/-}$ mice lowered spontaneous apoptosis of the thymocytes. Our result of activation-induced cell death by anti-CD3ε antibody in vitro was consistent with that of the spontaneous apoptosis of the animal in vivo. To compare the roles of CD3ε and CD3ζ in the induction of thymocyte apoptosis, the CD3$\varepsilon^{+/+}$ mice were crossed with CD3$\zeta^{-/-}$ mice. Not as expected, it was found that there was no compensatory effects on AICD of the thymocytes from their offspring (E600 x Z$^{-/-}$). AICD of thymocytes of E600 x Z$^{-/-}$ mice was similar to that of CD3$\zeta^{-/-}$ mice induced by either 2C11 or DEX (Fig. 2B).

From the preliminary results mentioned above, we come to the following conclusions, i.e., (1). CD3ε overexpression could block thymus development in the fetal transgenic mice and facilitate spontaneous apoptosis of thymocytes in young transgenic mice; (2). Thymocytes of the CD3$\varepsilon^{+/+}$ young transgenic mice were more sensitive to apoptosis than those of the wild type mice induced by anti-CD3ε or DEX; (3). CD3ζ inactivation could block thymus development in the fetal transgenic mice and render the remaining thymocytes to resist apoptosis induced by anti-CD3ε or DEX.

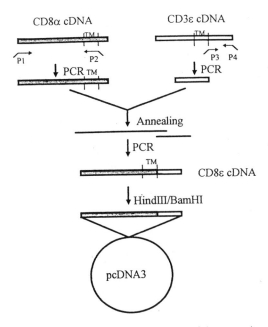

Figure 3. Cloning strategy of the chimera cDNA and the expression vector.

3. THE MOLECULAR MECHANISM OF APOPTOSIS MEDIATED BY CD3ε

With great interests in the above findings, we tried to design experiments to answer what is the molecular mechanism of lymphocyte apoptosis induced by CD3ε overexpression and does the apoptotic signaling occur through the ITAM in the CD3ε cytoplasmic tail? To address this question directly, a chimera by fusing the extracellular and transmembrane domains of human CD8α to the cytoplasmic domain of CD3ε was constructed (termed as CD8-ε) (Fig. 3). And the mutants of Y180 to F180 (CD8-ε Y180F) and Y171/Y180 to F171/F180 (CD8-ε Y171F Y180F) in the ITAM of CD3ε cytoplasmic tail were made from the CD8-ε construct. Expression vectors were made by cloning technology of these cDNA fragments into pcDNA3 plasmid. We assumed that mutation of tyrosine to phenoalanine might result in a block of the apoptotic signal transduction pathway through the ITAM in the CD3ε cytoplasmic tail based on the experimental evidence that the two tyrosines are necessarily phosphorylated during the activation process. The CD8α was chosen since we are sure that human Jurkat T cells are CD8α–negative and so that these cells could be used as an ideal host, thus the endogenous CD3ε could be ignored when stimulation of the transfectants was carried out by using anti-CD8α monoclonal antibody. The stable expression cells, termed as JK/CD8-ε, JK/CD8-ε Y180F and JK/CD8-ε Y171F Y180F respectively, were identified by staining the G418 resistant cells with FITC-labeled anti-human CD8α monoclonal antibody (PharMingen) and analyzed by flow cytometry (Coulter 752). Induction of apoptosis of the transfectants was carried out by stimulating the cells in plates with wells precoated with mouse anti-human CD8α monoclonal antibody (IgG1, 100μg/ml in PBS, PharMingen). The cells in wells uncoated with antibody were as the negative control and the cells transfected with

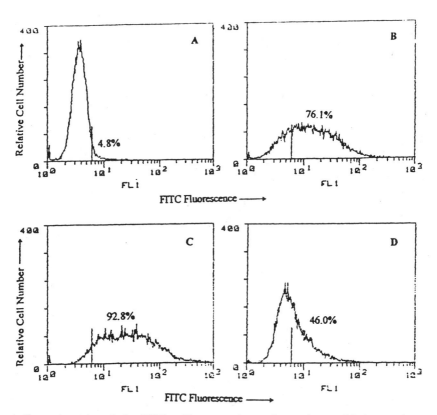

Figure 4. Flow cytometric analysis of CD8-ε chimera proteins on the membranes of Jurkat transfectants. Cells were stained with FITC-conjugated mouse anti-human CD8α mAb. A: pcDNA3 transfentant; B: CD8-ε/pcDNA3 transfentant; C: CD8-ε Y180F/pcDNA3 transfentant; D: CD8-ε Y171F Y180F/pcDNA3 transfentant.

pcDNA3 plasmid DNA only in the wells coated with antibody were as mock experiment. Percentages of the apoptotic cells were detected by PI exclusion and flow cytometry.

The flurocytometry analysis showed that untransfected Jurkat cells (Fig. 4A) were FITC-CD8α negative. The transfectants of CD8-ε, CD8-ε Y171F and CD8-ε Y171F Y180F were FITC-CD8α positive (Fig. 4B, C, D) respectively, indicating that the stable expression cell lines were established. Induction of the transfected cells with anti-CD8α mAb for 40 hrs, cell numbers of activation-induced cell death of the JK/CD8-ε were 23% and no apoptotic cells were found in the JK/CD8-ε Y180F and JK/CD8-ε Y171F Y180F cells detected by PI exclusion under the same conditions (Fig. 5). This clear-cut result provided for the first time an evidence that the ITAM in the cytoplasmic tail of CD3ε could relay and transduce the death signals from the CD8α stimulation to mediate apoptosis of human leukemia T lymphocytes, in which two tyrosine residues are equally important in the apoptotic signal transduction as well as activation processes.

The structure complexity of TCR/CD3 complex has previously hampered analysis of its structure and function relationship by a simple technology. This study provided a model for the study the molecular mechanism on a single amino acid residue basis in the CD3ε ITAM and demonstrated its possibility to pinpoint the change of the particular molecule in question in the intracellular death signal transduction events.

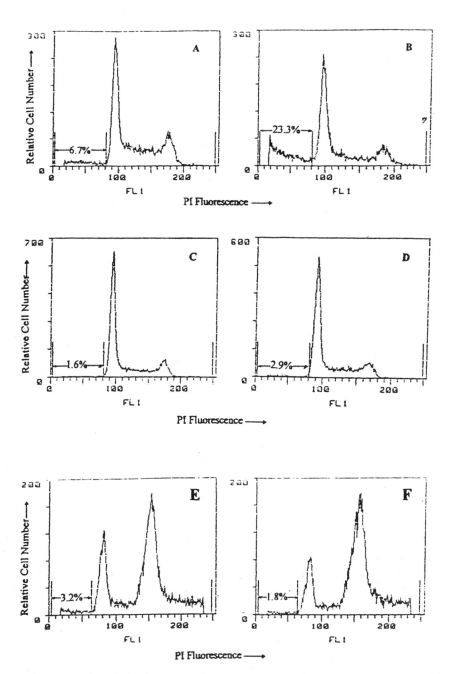

Figure 5. Flow cytometric analysis of apoptosis of the stable Jurkat transfectants induced by mouse anti-human CD8α mAb (IgG) for 40 hrs at 37¡æ. A: JK/CD8-ε cells. B: JK/CD8-ε cells stimulated with anti-CD8α mAb. C: JK/CD8-ε Y180F cells. D: JK/CD8-ε Y180F cells stimulated with anti-CD8α mAb. E: JK/CD8-ε Y171F Y180F cells. F: JK/CD8-ε Y171F Y180F cells stimulated with anti-CD8α mAb.

REFERENCES

1. Phillips J. H., Hori T., Nagler A., Bhat N., Spits H. and Lanier L. L., (1992) Ontogeny of human natural killer (NK) cells: fetal NK cells mediate cytolytic function and express cytoplasmic CD3ε, δ protein, J. Exp. Med. 175:1055–1066.
2. Baniyash M., Garcia-Morales P., Bonifacino J. S., Samelson L. E. and Klausner R. D., (1988) Disulfide linkage of the ζ and η chains of the T-cell receptor, J. Biol. Chem. 263:9874–9878
3. Jaime Sancho, Rafael Franco, Talal Chatila, Craig Hall and Cox Terhorst, (1993) The T cell receptor associated CD3-ε protein is phosphorylated upon T cell activation in the two tyrosine residues if a conserved signal transduction motif, Eur. J. Immunol., 23:1636–1642.
4. Mladen Mercep, Allan M. Weissman, Stuart J. Frank, Richard D. Klausener, Jonathan D. Ashwell, (1989) Activation-driven programmed cell death and T cell receptor ζη expression, Science, 246:1162–1165.
5. Behazine Combadiere, Matthew Freedman, Lina Chen, Elizabeth W. Shores, Paul Love and Michael J. Lenardo (1996) Qualitative and quantitative contributions of the T cell receptor ζ chain to mature T cell apoptosis, J. Exp. Med. 183:2109–2117.
6. Baoping Wang, Levelt C., She J., Dexian Zheng et al., CD3ε Overexpression Blocked the Thymus Development, Int. Immunol. 7:435–439, 1995.

12

MITOGENIC CYTOKINES PROMOTE APOPTOSIS

Possible Roles in Cellular Homeostasis

Yufang Shi,[1] Gordon B. Mills,[2] and Ruoxiang Wang[1]

[1]Department of Immunology, Holland Laboratory
American Red Cross
15601 Crabbs Branch Way
Rockville, Maryland 20855
[2]Department of Molecular Oncology
MD Anderson Cancer Center
University of Texas
1515 Holcombe Boulevard
Houston, Texas 77030

ABSTRACT

Successful cell proliferation relies on the activation of genes involved in cell cycle progression and concurrent inhibition of the function of tumor suppressor genes. In addition, there is a need to maintain the viability in proliferating cells. In the absence of cell viability factors cell cycle progression is intrinsically linked to programmed cell death through apoptosis. Whereas much attention has been focused on the molecular description of cell cycle machinery and the apoptotic pathway, less attention has been paid to the mechanisms for maintenance of the cell viability while cells are progressing through the cell cycle. The interaction of mitogenic cytokines with their cognate receptors promotes entry to the S-phase of cell cycle. It has also been suggested that cytokines could directly provide cell survival signals based on the observations that cytokine withdrawal leads to apoptosis. However, it is not clear whether the mitogenic activities and the survival effects of these cytokines are exerted by a single biochemical mechanism or carried out through separate signaling pathways. It is, thus, important to establish whether the survival effects of cytokines are achieved directly or by synergizing with the action of survival factors present in the environment such as interstitial fluid or blood *in vivo* or culture medium *in vitro*. We have addressed this question by stimulating cells with cytokines in the presence or absence of exogenous serum *in vitro*. When incubated with cytokines in serum-free medium, several cell types underwent apoptosis, a phenomenon called cytokine-promoted

Programmed Cell Death, edited by Shi *et al.*
Plenum Press, New York, 1997

apoptosis. Interestingly, we found that serum is the principle activator of the cell survival gene, *Bcl*-2; while mitogenic cytokines increase the expression of *Bax*. In this review, we discuss the relevance of cytokine-promoted apoptosis to the maintenance of cellular homeostasis during development and tumorigenesis.

Apoptosis, the main mechanism of programmed cell death, is a gene-directed process responsible for the elimination of excessive cells during development and detrimental cell types in pathophysiological situations (Wyllie *et al.*, 1980). Aberrant apoptosis has been implicated in the etiology of cancer and the process of autoimmunity (Hale *et al.*, 1996). It is also involved in the progression of some degenerative diseases as well as the evolution of drug resistance in tumors (Hickman, 1992). Deregulated expression of genes such as *Bcl-2*, loss of p53 expression, and autocrine activation of either apoptotic or anti-apoptotic signal transduction pathways may all contribute to pathological apoptosis (McConkey *et al.*, 1996).

Cell proliferation requires the modulation of at least two sets of genes: those that activate mitosis and those that maintain cell survival. The modulation of these genes is required to terminate cell cycle brakes, to activate cell cycle machinery, and to promote cell survival (Meikrantz and Schlegel, 1995). Much is known about the mechanisms of cell cycle promotion, the signals that regulate cell viability during cell cycle are yet to be determined.

Protooncogenes and tumor suppressor genes control and regulate cell cycle progression. In addition, both oncogenes and tumor suppressor genes play fundamental roles in the regulation of the pathways leading to programmed cell death (Chiarugi and Ruggiero, 1996). We have shown that the protooncogene c-*Myc* is required for activation-induced apoptosis in T cell hybridomas (Shi *et al.*, 1992). Ectopic expression of *Myc* enhances cell cycle progression and, at the same time, promotes apoptosis of the cycling cells (Evan *et al.*, 1992; Askew *et al.*, 1993). It has also been shown that cyclin A and the CDK kinase inhibitor, p27, play fundamental roles in anti-IgM-induced apoptosis in B-cell lymphomas (Ezhevsky *et al.*, 1996). This strongly suggests that cell cycle progression and apoptosis are tightly correlated, a notion which could have important implications in the understanding of the basic mechanisms regulating apoptosis. In this review, we explore the relationship between apoptosis and cell cycle progression by focusing on the recent studies in cytokine-promoted apoptosis, which will be important for the elucidation of the mechanisms by which normal cells manage to proliferate, and transformed cells undergo aberrant growth during tumorigenesis.

CELL CYCLE AND APOPTOSIS

Cell cycle regulation relies on a network of coordinated phosphorylation and dephosphorylation of Rb, p53, E2F family members, cyclins, and cyclin-dependent kinases. These components function coordinately to allow the progression of cell cycle (MacLachlan, 1994; Hinds and Weinberg, 1994). Exogenous growth stimulators (*e.g.* hormones, growth factors, and cytokines) trigger an intracellular kinase cascade, which in turn increases the expression of early nuclear proteins and consequently leads to successful cell cycle transit.

Paradoxically, many of these early response genes are also involved in the regulation of the cell death program (Meikrantz and Schlegel, 1995). In many investigation models, the location of each of these molecules in the cell cycle regulation cascade corresponds to its importance in determining the susceptibility of the cell to apoptosis (Chiarugi *et al.*, 1994). The expression of various cyclins and the activation of CDKs have been shown to be correlated to the onset of apoptosis. In many cell types, early nuclear proteins including c-*Myc*, c-*Jun,* c-*Fos*, and *cdc2* are induced after the initiation of the apoptosis pathway, about the same time when the phosphorylated form of the RB protein appears (Pandey and Wang, 1995).

Castration initiates extensive apoptosis of the secretory epithelial cells lining the ducts of the rat ventral prostate. The induction of apoptosis in these epithelial cells is preceded by the expression of *c-Fos* and *c-Myc* genes and the incorporation of bromodeoxyuridine into high-molecular-weight nuclear DNA prior to chromosomal DNA fragmentation (Colombel *et al.*, 1992). In the majority of situations, en route to apoptosis, cells are engaged in events typical of early cell-cycle traverse, including expression of genes that are intrinsic to the early G_1 phase of the cell cycle (Chiarugi *et al.*, 1994.; Meikrantz and Schlegel,1995). Nonetheless, deregulated traverse through the G_1 phase seems to be an abortive event, since these cells die shortly afterwards.

It is important to note that cells at specific stages of cell cycle are more susceptible to apoptosis. This could be due to the fact that at these stages cells require a higher threshold of cell survival signals. These stages could be considered as checkpoints for cell cycle progression. In clinical practice, it has been shown that cancer cells in cell cycle are more sensitive to a number of different chemotherapeutic agents (Kondo, 1995; Dubrez *et al.*, 1995). It has also been demonstrated that in serum deprivation-induced apoptosis, cells are first engaged in the entry of cell cycle before committing to cell death (Pandey and Wang, 1995). These observations, again, suggest that the pathways regulating cell proliferation and cell death are tightly coupled. Confirmation of this coupling may have dramatic implications for the understanding of the role of oncogenes in carcinogenesis. The molecular basis which determines whether a cell proceed through cell cycle or enter cell death pathway remains elusive. Our model system of cytokine-promoted apoptosis under serum-deprived conditions could provide a unique opportunity to initiate an identification of these determinants.

CYTOKINE-PROMOTED APOPTOSIS

Mitogenicity has been considered as a fundamental property of many cytokines. Growth factors and mitogenic cytokines induce cell cycle progression in responsive cells through specific activation of a series of immediate early response genes (Zola, 1996). Many growth factors and cytokines with the capability to regulate cell survival have been identified. In addition, signaling via several cell surface molecules, such as CD28 in our studies (Shi *et al.*, 1995; Radvanyi *et al.*, 1996) as well as CD40 in B cells (Choi *et al.*, 1995), was also shown to be able to enhance cell viability. Since many of these factors were identified within complex cell culture systems such as serum supplemented medium, it is not clear whether growth factors and cytokines promote cell viability directly, or act as co-factors with other mediators present in the culture environment.

In the absence of cell survival factors, such as in serum-deprived medium, overexpression of *c-Myc* (Evan *et al.*, 1992) or E2F-1 (Shan and Lee, 1994) in fibroblasts resulted in both enhanced mitogenesis and enhanced apoptosis. Whereas *c-Myc* promotes apoptosis in serum-free medium, addition of IGF-1 was sufficient to counteract the apoptotic activity and to maintain cell viability as well as cell cycle progression (Harrington *et al.*, 1994). These findings strongly indicate that successful cell proliferation requires active promotion of both mitogenecity and cell viability.

The mytogenic cytokine interleukin-2 (IL-2) is a primary T lymphocyte growth factor. At the same time, IL-2 also participates in the initiation of T cell apoptosis. Lenardo

(1991) and Russell *et al.* (1991) reported that IL-2 was required for activation-induced apoptosis in primary activated mature murine T cells. Ucker *et al.* (1991) proposed that the apoptosis signals provided by IL-2 were linked to cell cycle promotion. We have recently shown (Radvanyi *et al.*, 1996) that rapamycin, which did not block activation-induced apoptosis in T-cell hybridomas, blocked IL-2-dependent cell cycle progression and reactivation-induced apoptosis in primary T cells. This further supports the role of IL-2 promoted cell cycling in the pathway leading to activation-induced apoptosis in activated T cells. In addition, we have found that at the same time when T cell hybridomas ceased to proliferate under serum-free conditions, they become resistant to activation-induced apoptosis (Shi *et al.*, data not shown).

Under certain conditions, many cytokines seem to be able to induce apoptosis directly. For example, TNF was identified by its ability to induce cell death in some tumor cell lines. This death-induction property, however, was highly restricted to target cells under specific conditions, *i.e.*, they needed to be treated with metabolic inhibitors or under serum-free conditions, under which cell survival signals were depleted (Higuchi *et al.*, 1995). In the presence of survival signals, TNF, like many other cytokines, functions as a mitogenic signal in many cell lines (Borset *et al.*, 1996; Branch and Guilbert, 1996). Furthermore, recent experiments have demonstrated that TNF protects bone marrow stem cells and hair follicular cells from chemotherapy-induced apoptosis (Wong, 1995; Tiwari *et al.*, 1991, and Grace Wong, presented at this meeting). Thus, TNF has the ability to promote or inhibit apoptosis depending on intrinsic cellular characteristics as well as on the extracellular milieu.

NGF is the major factor regulating neuronal proliferation. Paradoxically, when neuronal cells were cultured in serum-free media, NGF became a potent apoptosis-inducing factor (Rabizadeh *et al.*, 1993). Kim *et al.* (1996) recently reported that mitogenic factors PDGF or EGF could induce a kidney fibroblast cell line to undergo apoptosis. Combination of PDGF and EGF exhibited synergistic effect in inducing apoptosis in this cell line. This apoptosis could be blocked by the addition of IGF-1 to the tissue culture medium. Our recent experiments demonstrated that growth factors including IL-2, IL-3, EGF, and FGF could induce apoptosis in responsive cells in serum-free medium. This apoptosis could be effectively suppressed by transferrin (data not shown) or some lipids such as lysophosphatidic acid (Shi *et al.*, submitted). It seems that growth factors may promote successful cell cycle progression only when sufficient cell survival signals are present. Without survival signals, cell cycling would become abortive with cells entering the apoptotic pathway.

PROTOONCOGENES AND APOPTOSIS

Protooncogenes can be categorized into two groups, based on their ability to influence cell proliferation and apoptosis (Hale *et al.*, 1996). One group of the oncogenes activate cells to exit from a growth-arrest state into a state in which both apoptosis and entry to S-phase becomes possible (Liebermann *et al.*, 1995). Activated cells are often more sensitive to apoptosis induced by a wide variety of agents, including several drugs used in cancer chemotherapy. The choice between cell death and continued proliferation appears to be determined by second signals mediated by another group of oncogenes that mediate cell survival. The *c-Myc* gene belongs to the first group, whereas the genes for *Bcl-x*, *Bcl-2*, and *v-Abl* belong to the second group. This functional classification may facilitate the characterization of oncogenes.

One example is *c-Neu,* which belongs to the first group of oncogenes. Constitutive expression of the activated form of *c-Neu* resulted in some signs of the transformation of the breast epithelial cell line HB4a (Harris *et al.,* 1995). These transformed cells, however, were not able to give rise to tumors in nude mice, indicating that the expression of *c-Neu* alone was insufficient to cause tumorigenic progression to a full malignant phenotype. On the other hand, under conditions of serum deprivation, the *c-Neu* transfected cells undergo apoptotic cell death (Harris *et al.,* 1995). Thus, the prevention of oncogene-driven apoptosis is probably important for tumorigenesis. It seems that in order to become a tumor, a cell needs to constitutively express survival genes. This also indicates that tumorigenesis is a process of deregulation of multiple genes involved in both proliferation and survival. Indeed, genetic manipulation of a single oncogene does not always results in tumor formation. As an example, c-*Myc* itself is not sufficient to transform 3T3 cells. In the presence of an activated Ras, however, c-*Myc* would become an efficient transforming gene (Murray *et al.,* 1983). Thus, in the progression to malignancy, cells constitutively express oncogenes that function both in cell cycle progression and in cell survival. Tumorigenesis is, therefore, a result of the deregulation of multiple genes that are involved in both cell proliferation and cell survival.

The *c-Myc* protooncogene is a critical early response gene for the stimulation of proliferation initiated by a number of growth factors (Lee, 1989). It has long been implicated as a critical element in the regulation of normal cell growth, being required for transition through the G_1 to S phase in non-transformed cells (Chiarugi *et al.,* 1994). Consistent with its role as a central regulator of cell growth, deregulated expression of *c-Myc* serves as a potent mitogenic stimulus and is associated with the development of an array of tumor types (Garte, 1993, Spencer and Groudine, 1991). The expression of *c-Myc* is indeed upregulated in a number of tumor cell lineages (Alitalo and Schwab, 1986).

In addition to promoting cell cycle progression, *c-Myc* plays a role in promoting apoptosis. As stated above, we have demonstrated that antisense oligodexoynucleotides could inhibit *c-Myc* expression and prevent activation-induced apoptosis in T cell hybridomas (Shi *et al.,* 1992). Furthermore, while *c-Myc*-transfected fibroblasts exhibited enhanced proliferation in the presence of serum, they underwent rapid apoptosis upon deprivation of serum (Evan *et al.,* 1992). Thus, the decision of *c-Myc*-expressing cells to proliferate or to undergo apoptosis is likely modulated by the availability of cell survival factors, which may be different from cell to cell.

The involvement of *c-Myc* in the promotion of apoptosis is also seen in other systems. Cell death induced by prostaglandin E2 in human lymphocytes, as well as TNF-induced apoptosis in HeLa cells (Janicke *et al.,* 1994) and 3T3 fibroblasts (Klefstrom *et al.,* 1994) are *c-Myc*-dependent. However, the molecular mechanism(s) by which *c-Myc* causes cells to undergo apoptosis is unknown. It may simply function to induce cells to progress through the early G_1, a stage where cells enter apoptosis unless additional survival factors are provided. On the other hand, *c-Myc* may directly induce the expression of pro-apoptotic proteins which subsequently induce cells to initiate the suicide response.

The functionality of *c-Myc* as a specific transcription activator and transforming protein is based on its dimerization with the regulatory protein *Max,* which is a obligate heterodimeric partner of c-*Myc* (Wu *et al.,* 1996). Efficiency of the heterodimerization between c-*Myc* and *Max* is regulated by *Mad* and *Mxi1* proteins. These two molecules have the capability to heterodimerize with *Max,* and so to compete with c-*Myc* for *Max* (Hurlin *et al.,* 1994). Alterations in the levels of *Max*-associated proteins such as *Mxi1* can modulate the level of functional *Myc/Max* complexes. Overexpression of *Mxi1* inhibits *Myc*-dependent gene expression in a dose-dependent manner both *in vivo* and *in vitro*

(Larsson *et al.*, 1993). This inhibition also alters whether a cell proliferates or dies *via* apoptosis (Wu *et al.,* 1996; Chen *et al.,* 1995). The *Myc/Max* heterodimer is thought to be able to regulate at least two subsets of genes: those required for cell proliferation and those for apoptosis. The action of these genes can be distinguished at the molecular level by transfection with cell survival genes (e.g. *Bcl*-2), which would counteract the effects of the apoptosis genes. It remains unclear whether these subsets are mutually exclusive or overlapping, since only a limited number of *Myc*-regulated target genes have been identified and their role in growth or apoptosis have not yet been fully defined. A recent report showed that *cdc25A*, an integral component of the cell cycle machinery, could also induce apoptosis by acting downstream of c-*Myc* (Galaktionov *et al.*, 1996).

p53 AND p21

The tumor suppressor p53 plays a major role in mediating G_1 arrest in many situations, for example, in response to DNA damage (Montenarh, 1992). It has a dualistic function regulating cellular commitment to apoptosis as well as suppression of cell transformation. Also, it is the most frequently affected gene detected in various human cancers. P53 guards genome integrity by regulating the transcription of a series of other regulatory genes, such as p21, GADD45, MyD118 and mdm2 (Vairapandi, *et al.*, 1996). Through control of the expression of these genes, p53 is involved in DNA replication and in DNA repair, as well as in cell-cycle control and apoptosis. P53 has been shown to downregulate *Bcl-2* expression (Miyashita *et al.*, 1994). Importantly, the apoptosis enhancing gene *Bax* has been shown to be the immediate early response gene induced by p53 (Miyashita and Reed, 1995). Deregulated expression of wild-type p53 (wt-p53) protein induced apoptosis. Interestingly, serum deprivation could induce p53-dependent apoptosis. P53 has also been shown to promote apoptosis induced by the depletion of specific growth factors (Blandino *et al.*, 1995), suggesting that p53 might be involved in response to growth factor deprivation, which is currently being examined in our model for cytokine promoted apoptosis.

The cyclin-dependent kinase inhibitor p21WAF1/Cip1 (p21) could inhibit cellular proliferation through a G_1-phase checkpoint. It has been identified as a downstream mediator of p53 and is required for p53-induced G_1 arrest (Gartel *et al.*, 1996). Antisense oligonucleotides to p21WAF1 mRNA partially block the expression of p21WAF1 and also inhibit programmed cell death. It was suggested that p21 exerted its effect directly or indirectly by inhibiting the function of another regulatory protein, E2F-1 (Dimri *et al.*, 1996). It is, therefore, also intriguing to examine the role of both p21 and E2F-1 in our model of the cytokine promoted apoptosis.

E2F AND Rb

The E2F family of transcription factors is regulated during the cell cycle through interactions with Rb and Rb-related proteins (Ewen, 1994). Phosphorylation or sequestration of Rb by viral oncoproteins can release E2F from the E2F-Rb complex (Nevins, 1992). The E2F DNA-binding complex consists of a heterodimer of E2F-1 and the constitutively expressed DP family proteins (Martin *et al.*, 1995). This heterodimerization is required for further association of E2F proteins with pRb and the pRb-related proteins p107 and p130. The expression of E2F-1 is sufficient to induce fibroblasts to exit G_0 phase and enter S phase (Johnson *et al.*, 1993). Commitment of mammalian cells to enter S phase enables the transcription factor E2F-1 to activate certain genes, whose products contribute to cell cycle progression. On the other hand, overexpression of E2F-1 in fibroblasts induces S-phase entry and promoted apoptosis (Shan and Lee, 1994). In IL-3-dependent 32D myeloid cells, in the absence of IL-3, E2F-1 alone was sufficient to induce apoptosis. This is similar to the observations in c-*Myc* transfected 32D cells (Hiebert *et al.*, 1995). In contrast, DP-1, the other component of the E2F DNA-binding complex, was not sufficient to induce cell cycle progression or to alter the rate of apoptosis following IL-3 depletion.

The product of the retinoblastoma tumor-suppresser gene (*Rb*) is ubiquitously expressed. Rb protein negatively regulates the transition of cells from the G_1 phase to the S phase of the cell cycle. In terms of molecular mechanism, the growth-inhibitory effects of pRb are exerted, at least in part, through its interaction with the E2F family of transcription factors (Newins, 1992). Thus, E2F plays a role in the regulation of both cell cycle progression and cell death. In a system with tetracycline-inducible dominant-negative expression, mutated E2F could efficiently protect c-*Myc* transfected fibroblasts from serum starvation-induced apoptosis. The protective effect lasted for a period of 10 days, whereas the control cells began to die after 24 hours after starvation. We have shown that even under serum-deprived conditions, cytokines induce the expression of *Myc* (data not shown). It appears that E2F family members are also involved in the cytokine promoted apoptosis.

Bcl-2 FAMILY

The protooncogene *Bcl-2* is highly expressed in many t(14:18)-containing human follicular lymphomas (Korsmyer, 1992). This translocation results in the recombination of the *Bcl-2* gene from chromosome 18 with the transcriptionally active immunoglobulin locus on chromosome 14. High level expression of *Bcl-2* in folicular lymphoma cells was thought to contribute to the tumorigenesis. *Bcl-2* inhibits apoptosis in many cell types following a variety of stimuli, including growth factor withdrawal, irradiation, some chemotherapeutic drugs, and high level c-*Myc* expression (Reed, 1994). It may form a homodimer or may form a heterodimer with related proteins, such as *Bax*. High expression of *Bax* could accelerates apoptosis following growth factor withdrawal. It is proposed that *Bax* augments apoptosis through dimerization with *Bcl*-2, preventing *Bcl*-2 homodimer from acting as cell survival signals.

Several proteins homologous to *Bcl*-2 and *Bax* have thus far been identified (Craig, 1995). The proteins of the *Bcl*-2 family appear to play a major role in regulating programmed cell death. These proteins interact with each other through a complex dimerization network and the fine balance of the various dimers possibly determines the cell viability. The *Bcl*-2, *Bcl*-x, and *Mcl*-1 proteins from this family may protect cells from cell death, whereas other members of this family, such as *Bax*, *Bak*, and *Bad* appear to promote cell death (Craig, 1995). Currently, the regulatory mechanism for expression of these genes has not been elucidated, nor the molecular mechanism by which their proteins regulate cell death. It appears that the members of the *Bcl*-2 family act, both in the form of homodimers and heterodimers, to shunt cells either into cell division or into apoptosis. Understanding the mechanisms controlling the balance of cell cycling versus cell death would cast insights into new approaches to the control of abnormal cell growth and tumorigenesis.

Overall, there exist two separate sets of regulatory components required for cell proliferation. The coordination of mitogenic and survival signals ensures effective cell cycling. A lack of cell survival signals may lead to an abortive cell cycle and ultimately to apoptosis (summarized in Fig.1).

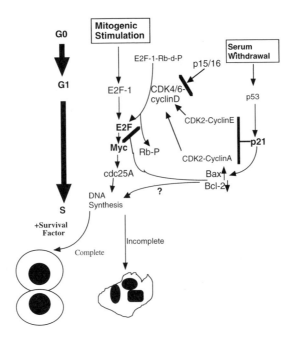

Figure 1. The link between cell proliferation and apoptosis. CDK inhibitor p21 is a target of p53. p53 binds the p21 promoter and activates the transcription of p21. Consequently, cell cycle progression is blocked at the G_1 phase through the suppression of CDK activity. pRb is a substrate of a cyclinD-CKD4/6 complex. pRb functions to suppress cell cycle progression at the G_1 phase associated with the E2F transcription factor. Phosphorylation of pRb by a cyclinD-CDK4/6 complex releases active E2F, which promotes the expression of genes, such as *c-Myc*, whose products may play a crucial role in controlling G_1–S progression. In addition, p21 can also directly inhibit the transcription of E2F and probably directly interfere with DNA synthesis and ultimately lead to apoptosis. The hypothesis is that continuous cell proliferation requires a balance between *Myc* expression induced by cytokine stimulation, and p53 expression inhibited by growth factors in serum.

CYTOKINES INDUCE APOPTOSIS IN THE ABSENCE OF SERUM

Our recent study showed that in BAF/BO3 cells transfected with IL-2 receptor β-chain, mitogenic cytokines IL-2 and IL-3 could increase the expression of both c-*Myc* and *Bax*, regardless of the presence or absence of serum. Although BAF/BO3 cells proliferated upon stimulation with IL-3 in the presence of serum (data not shown), they undergo typical apoptosis when stimulated by IL-3 in the absence of serum. This cytokine-induced increased rate of apoptosis can also be observed in EGF stimulated Rat1 fibroblasts, similar to that observed by Kim *et al.* (1996). Therefore, cytokine-promoted apoptosis may represent a general biological phenomenon.

OVEREXPRESSION OF *Bcl-2* PREVENTS CYTOKINE-INDUCED CELL DEATH

We have demonstrated that serum is required for the optimal expression of *Bcl-2* while cytokines enhance the expression of *Bax* in the presence or absence of serum (manuscript submitted). Thus, cytokine-induced cell death is due to an imbalance between *Bax* and *Bcl-2* levels, and artificially increasing *Bcl-2* levels should decrease or abrogate

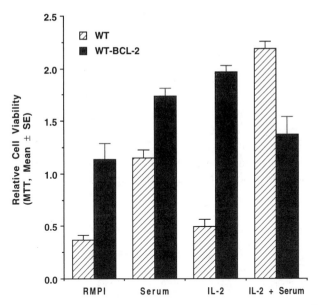

Figure 2. Bcl-2 overexpression prevents cytokine-induced apoptosis in BAF/BO3 cells. BAF/BO3 cells expressing the wild type IL-2Rβ (WT) were infected with a retrovirus containing Bcl-2 and subjected to puromycin selection. Cells were plated in RPMI in 96-well plates at 2×10^4 cells per well with recombinant human IL-2 at 100 U/ml in the presence or absence of serum for 40 hour. Cell viability was assessed by MTT staining.

the ability of cytokines to induce cell death. To determine the effect of constitutive *Bcl*-2 expression in BO3 cells on cytokine-induced cell death, a previously characterized *Bcl*-2 expressing cell line created by infection with a *Bcl*-2 containing retrovirus was evaluated. Expression of *Bcl*-2 conferred resistance to cytokine-induced apoptosis (Fig. 2).

ISOLATION OF CELL SURVIVAL GENES

We have made a cDNA from logarithmically dividing BAF/BO3 growing in 10% serum plus IL-3. The cDNA was cloned unidirectionaly into the ZAP-Express vector (Stratagene) carrying the Neo selection marker. Phagemid DNA representing all the clones in the library were isolated by autoexcision. The phagemids were transfected into BO3 cells by hypersonic pinocytosis, and transfectants then subjected to stimulation with IL-3 in the absence of serum for 5 days. The surviving cells were then subjected to G418 selection. Surviving cells were cloned and the transfected cDNAs was isolated from genomic DNA by PCR with primers derived from sequences in the plasmid. One of several isolated clones, SRG-1, is about 1.0 kb in length and sequencing from both ends shows it contains a complete ORF. It is a novel gene contains tentative proline and glycine-rich collagen domains with some homology to a uncharacterized viral protein. The mechanism by which this ORF regulates apoptosis is unknown.

CONCLUSION

The effects of signals initiated by the interaction between cytokines and their receptors include the promotion of cell cycle progression, induction of cell differentiation, maintenance of cell viability, and the induction of cell death. In order to maintain cellular homeostasis of multicellular organisms, cytokine receptors may have been selected during evolution not only to promote cell viability and proliferation, but also to direct cells into apoptosis under certain circumstances. As a corollary, the coupling of the mitogenic and apoptotic pathways may ensure that proliferation occurs only when there is a balance among the proliferative and apoptotic effects of factors and stimuli to which a cell is exposed. In the event of constitutive activation of mitogenic pathways, a cancer cell can only persist and proliferate if there is a concurrent induction of survival signals such as those mediated by the members of the *Bcl*-2 family. Thus, the studies of the mechanisms by which cytokines promote apoptosis under certain circumstances may provide fundamental information for the general understanding of cellular homeostasis.

REFERENCES

Alitalo, K., and Schwab, M. (1986). Oncogene amplification in tumor cells. Adv Cancer Res 47, 235–281.

Askew, D.S., Ihle, J.N., and Cleveland, J.L. (1993). Activation of apoptosis associated with enforced *Myc* expression in myeloid progenitor cells is dominant to the suppression of apoptosis by interleukin-3 or erythropoietin. Blood 82, 2079–2087.

Blandino, G., Scardigli, R., Rizzo, M.G., Crescenzi, M., Soddu, S., and Sacchi, A. (1995). Wild-type p53 modulates apoptosis of normal, IL-3 deprived, hematopoietic cells. Oncogene 10, 731–737.

Borset, M., Medvedev, A.E., Sundan, A., and Espevik T. (1996). The role of the two TNF receptors in proliferation, NF-kappa B activation and discrimination between TNF and LT alpha signalling in the human myeloma cell line OH-2. Cytokine 8, 430–438.

Branch, D.R., and Guilbert, L.J. (1996). Autocrine regulation of macrophage proliferation by tumor necrosis factor-alpha. Exp. Hematol. 24, 675–681.

Chiarugi, V., and Ruggiero, M. (1996). Role of three cancer "master genes" p53, *Bcl2* and c-*Myc* on the apoptotic process. Tumori. 82, 205–209.

Choi, M.S., Boise, L.H., Gottschalk, A.R., Quintans, J., Thompson, C.B., and Klaus, GG. (1995). The role of *Bcl*-XL in CD40-mediated rescue from anti-mu-induced apoptosis in WEHI-231 B lymphoma cells. Eur. J. Immunol. 25,1352–1357.

Colombel, M., Olsson, C.A., Ng, P.Y., and Buttyan, R. (1992). Hormone-regulated apoptosis results from reentry of differentiated prostate cells onto a defective cell cycle. Cancer Res. 52, 4313–4319.

Dimri, G.P., Nakanishi, M., Desprez, P.Y., Smith, J.R., and Campisi, J. (1996). Inhibition of E2F activity by the cyclin-dependent protein kinase inhibitor p21 in cells expressing or lacking a functional retinoblastoma protein. Mol. Cell. Biol. 16, 2987–2997.

Dubrez, L., Goldwasser, F., Genne, P., Pommier, Y., and Solary, E. (1995). The role of cell cycle regulation and apoptosis triggering in determining the sensitivity of leukemic cells to topoisomerase I and II inhibitors. Leukemia 9,1013–1024.

Evan, G.I., Wyllie, A.H., Gilbert, C.S., Littlewood, T.D., Land, H., Brooks, M. , Waters, C.M., Penn, L.Z., and Hancock, D.C. (1992). Induction of apoptosis in fibroblasts by c-*Myc* protein. Cell 69, 119–128.

Ewen, M.E. (1994). The cell cycle and the retinoblastoma protein family. Cancer Metastasis Rev. 13, 45–66.

Ezhevsky, S.A., Toyoshima, H., Hunter, T., and Scott, D.W. (1996). Role of cyclin A and p27 in anti-IgM induced G1 growth arrest of murine B-cell lymphomas. Mol. Biol. Cell 7, 553–564.

Galaktionov, K., Chen, X., and Beach, D. (1996). Cdc25 cell-cycle phosphatase as a target of c-*Myc*. Nature 382, 511–517.

Garte, S.J. (1993). The c-*Myc* oncogene in tumor progression. Crit Rev Oncogene 4, 435–449.

Gartel, A.L., Serfas, M.S., and Tyner, A.L. (1996). p21-negative regulator of the cell cycle. Proc. Soc. Exp. Biol. Med. 213, 138–149.

Hale, A.J., Smith, C.A., Sutherland, L.C., Stoneman, V.E., Longthorne, V.L., Culhane, A.C., and Williams, G.T. (1996). Apoptosis: molecular regulation of cell death. Eur. J. Biochem. 236,1–26.

Harrington, E.A., Bennett, M.R., Fanidi. A., and Evan, GI. (1994). c-*Myc*-induced apoptosis in fibroblasts is inhibited by specific cytokines. EMBO J 13, 3286–3295.

Harris, R.A., Hiles, I.D., Page, M.J., and O'Hare, M.J. (1995). The induction of apoptosis in human mammary luminal epithelial cells by expression of activated c-neu and its abrogation by glucocorticoids. Br. J. Cancer 72, 386–392.

Hickman, JA. (1992). Apoptosis induced by anticancer drugs. Cancer Metastasis Rev 11,121–139.

Hiebert, S.W., Packham, G., Strom, D.K., Haffner, R., Oren, M., Zambetti, G., and Cleveland, J.L. (1995). E2F-1:DP-1 induces p53 and overrides survival factors to trigger apoptosis. Mol. Cell. Biol. 15, 6864–6874.

Higuchi, M., Singh, S., and Aggarwal, B.B. (1995). Characterization of the apoptotic effects of human tumor necrosis factor: development of highly rapid and specific bioassay for human tumor necrosis factor and lymphotoxin using human target cells. J. Immunol. Methods 178, 173–181.

Hinds, P.W., and Weinberg, RA. (1994). Tumor suppressor genes. Curr. Opin. Genet. Dev. 4, 135–141.

Hurlin, P.J., Ayer, D.E., Grandori, C., and Eisenman, R.N. (1994). The Max transcription factor network: involvement of Mad in differentiation and an approach to identification of target genes. Cold Spring Harb. Symp. Quant. Biol. 59, 109–116.

Janicke, R.U., Lee, F.H., and Porter, A.G. (1994). Nuclear c-*Myc* plays an important role in the cytotoxicity of tumor necrosis factor alpha in tumor cells. Mol. Cell. Biol. 14, 5661–5670.

Johnson, D.G., Schwarz, J.K., Cress, W.D., and Nevins, J.R. (1993). Expression of transcription factor E2F1 induces quiescent cells to enter S phase. Nature 365, 349–352.

Klefstrom, J., Vastrik, I., Saksela, E., Valle, J., Eilers, M., and Alitalo, K. (1994). c-*Myc* induces cellular susceptibility to the cytotoxic action of TNF-alpha. EMBO J. 13, 5442–5450.

Kondo S. (1995). Apoptosis by antitumor agents and other factors in relation to cell cycle checkpoints. J. Radiat. Res. (Tokyo) 36, 56–62.

Larsson, L.G., Pettersson, M., Oberg, F., Nilsson, K., and Luscher, B. (1994). Expression of mad, mxi1, max and c-*Myc* during induced differentiation of hematopoietic cells: opposite regulation of mad and c-*Myc*. Oncogene 9, 1247–1252.

Lee, W.M. (1989). The *Myc* family of nuclear proto-oncogenes. Cancer Treat Res 47, 37–71.

Lenardo, M.J. (1991). Interleukin-2 programs mouse alpha beta T lymphocytes for apoptosis. Nature 353, 858–861.

Liebermann, D.A., Hoffman, B., and Steinman, R.A. (1995). Molecular controls of growth arrest and apoptosis: p53-dependent and independent pathways. Oncogene 11, 199–210.

MacLachlan, T.K., Sang, N., and Giordano, A. (1995). Cyclins, cyclin-dependent kinases and cdk inhibitors: implications in cell cycle control and cancer. Crit. Rev. Eukaryot. Gene Expr. 5,127–156.

Martin, K., Trouche, D., Hagemeier, C., and Kouzarides, T. (1995). Regulation of transcription by E2F1/DP1. J. Cell. Sci. Suppl. 19, 91–94.

McConkey, D.J., Zhivotovsky, B., and Orrenius, S. (1996). Apoptosis-molecular mechanisms and biomedical implications. Mol. Aspects Med. 17, 1–110.

Meikrantz, W, and Schlegel, R. (1995). Apoptosis and the cell cycle. J. Cell. Biochem. 58,160–174.

Miyashita, T., Harigai, M., Hanada, M., and Reed, J.C. (1994). Identification of a p53-dependent negative response element in the *Bcl*-2 gene. Cancer Res. 54, 3131–3135.

Miyashita, T., and Reed, J.C. (1995). Tumor suppressor p53 is a direct transcriptional activator of the human *Bax* gene. Cell 80, 293–299.

Montenarh, M. (1992). Biochemical, immunological, and functional aspects of the growth-suppressor/oncoprotein p53. Crit. Rev. Oncog. 3, 233–256.

Murray, M.J., Cunningham, J.M., Parada, L.F., Dautry, F., Lebowitz, P., and Weinberg, R.A. (1983). The HL-60 transforming sequence: a ras oncogene coexisting with altered *Myc* genes in hematopoietic tumors. Cell 33, 749–757.

Nevins, J.R. (1992). E2F: a link between the Rb tumor suppressor protein and viral oncoproteins. Science 258, 424–429.

Pandey, S., and Wang, E. (1995). Cells en route to apoptosis are characterized by the upregulation of c-fos, c-*Myc*, c-jun, cdc2, and RB phosphorylation, resembling events of early cell-cycle traverse. J. Cell. Biochem. 58, 135–150.

Pica, F., Franzese, O., D'Onofrio, C., Bonmassar, E., Favalli, C., and Garaci E. (1996). Prostaglandin E2 induces apoptosis in resting immature and mature human lymphocytes: a c-*Myc*-dependent and *Bcl*-2-independent associated pathway. J. Pharmacol. Exp. Ther. 277, 1793–1800.

Radvanyi, L.G., Shi, Y.F., Mills, G.B., and Miller, R. (1996). Cell cycle progression out of G1 sensitizes primary-cultured nontransformed T cells to TCR-mediated apoptosis. Cell. Immunol. 170: 260–273.

Rabizadeh, S., Oh, J., Zhong, L.T., Yang, J., Bitler, C.M., Butcher, L.L., and Bredesen, D.E. (1993). Induction of apoptosis by the low-affinity NGF receptor. Science 261, 345–348.

Russell, J.H., White, C.L., Loh, D.L., and Meleedy-Rey, P. (1991). Receptor-stimulated death pathway is opened by antigen in mature T cells. *Proc. Natl. Acad. Sci., U.S.A.*, 88, 2151–2155.

Shan, B., and Lee, W.H. (1994). Deregulated expression of E2F-1 induces S-phase entry and leads to apoptosis. Mol. Cell. Biol. 14, 8166–8173.

Shi, Y.S, Bissonnette, R.P, Glynn, J.M, Cotter, T.G., Guilbert, J.L., and Green, D.R. (1992). Inhibition of activation-induced apoptosis in T cell hybridomas by antisense oligodeoxynucleotides corresponding to *c-Myc*. Science 257, 212–214.

Shi, Y.S., Radvanyi, L.G., Mills, G.B., and Miller, R.G. (1995). CD28 ligation inhibits TCR-induced apoptosis during a primary T cell response and prevents induction of T cell hyporesponsiveness. J. Immunol. 156,1788- 1798.

Spencer, C.A., and Groudine, M. (1991). Control of c-*Myc* regulation in normal and neoplastic cells. Adv. Cancer Res. 56,1–48.

Tiwari, R.K., Wong, G.Y., Liu, J., Miller, D., and Osborne, M.P. (1991). .Augmentation of cytotoxicity using combinations of interferons (types I and II), tumor necrosis factor-alpha, and tamoxifen in MCF-7 cells. Cancer Lett. 61, 45–52.

Ucker, D.S. (1991). Death by suicide: one way to go in mammalian cellular development? New Biol. 3, 103–109.

Vairapandi, M., Balliet, A.G., Fornace, A.J. Jr, Hoffman, B., and Liebermann, D.A. (1996). The differentiation primary response gene MyD118, related to GADD45, encodes for a nuclear protein which interacts with PCNA and p21 WAF1/CIP1. Oncogene 12, 2579–2594.

Wong, G.H. Protective roles of cytokines against radiation: induction of mitochondrial MnSOD. Biochim. Biophys. Acta. 1271, 205–209.

Wu, S., Pena, A., Korcz, A., Soprano, D.R., and Soprano, K.J. (1996). Overexpression of Mxi1 inhibits the induction of the human ornithine decarboxylase gene by the *Myc*/Max protein complex. Oncogene 12, 621–629.

Wyllie, A.H., Kerr, J.F.R., and Currie, A.R. (1980). Cell death: The significance of apoptosis. Int. Rev. Cytol. 68, 251–306.

Zola H. (1996). Analysis of receptors for cytokines and growth factors in human disease. Dis. Markers 12, 225–240.

13

MOLECULAR MECHANISMS OF GROWTH AND DEATH CONTROL OF HEMATOPOIETIC CELLS BY CYTOKINES

Jeffrey J. Y. Yen,[1] Hsin-Fang Yang-Yen,[2] Huei-Mei Huang,[1] Yueh-Chun Hsieh,[1] Shern-Fwu Lee,[1] Jyh-Rong Chao,[2] and Jian-Chuan Lee[2]

[1]Institute of Biomedical Sciences
[2]Institute of Molecular Biology
Academia Sinica
Taipei, Taiwan

1. INTRODUCTION

In responding to the challenge of exogenous antigen(s), there is always a dramatic expansion of certain white blood cell populations in order to achieve tasks of self-defense. It is the cytokines secreted by various immunological cells which are responsible for provoking exponential proliferation and differentiation of these blood cells through binding to their own specific cell surface receptors. Not only the rate of proliferation, but also the length of life span of these blood cells increases. Usually some hundred-fold increase of a single cell population can be achieved within several days. However, once the antigen is eliminated from the body, a rapid clearance of these cell types can be expected within several days which returns the hematopoietic cell population to its normal size. This occurs in response to the decrease in the blood concentration of these essential cytokines. Both rapid expansion of desired cell lineage in emergency and rapid elimination of un-warranted cell population after recovery are characteristics of a successful immune response. Interference in the regulation of above mentioned control always results in immune-pathogenesis. Utilizing the factor-dependent cell line as a model system we are interested in the mechanisms of apoptosis initiation by cytokine deprivation, apoptosis suppression by cytokines, and possibly on the mechanisms underlying leukemogenesis.

2. CYTOKINE RECEPTOR ARCHITECTURE

The complexity of regulation in the immune response is partially reflected in the architecture of cytokine receptors. Most of the cytokine receptor complexes consist of multi-subunits and can be grouped into at least three sub-families according to the distinct common receptor subunit shared by members of each receptor sub-family (for review see

Programmed Cell Death, edited by Shi *et al.*
Plenum Press, New York, 1997

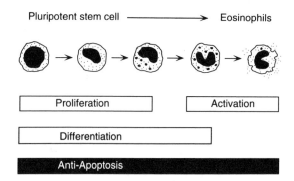

Figure 1. Functions of human interleukin 5 (hIL-5) in eosinophil differentiation and activation. Human IL-5 specifically drives the hematopoietic progenitor cells to differentiate into eosinophils via its surface receptor complexes. During this process several activities are transduced through the IL-5 receptors which include proliferation and differentiation of the progenitor cells and activation of degranulation of mature eosinophils. Recently, anti-apoptosis activity is described as an additional function of IL-5 which not only supports the viability of differentiating eosinophils but also determines the life span of mature eosinophils in peripheral blood.

Kishimoto, et al., 1994). In myeloid cytokine receptor family, including IL-3, GM-CSF, and IL-5, each cytokine has their own ligand-specific α subunit, and shares a signal transducer subunit, the receptor common β chain (βc). The α subunit is a low-affinity binding site for its cognate ligand and the βc subunit binds to ligand with high affinity when it co-exists with α subunit, although βc subunit does not bind to ligand by itself. Distinguishing from the α subunits in other receptor sub-families, IL-5Rα and GMRα subunits have both been demonstrated to be essential for functions of receptor complexes containing βc subunit (Sakamaki, et al., 1992; Takaki, et al., 1990). A proline-riched region in the cytoplasmic domain of these receptor α subunits, whose sequences are similar to the conserved Box I among cytoplasmic proximal regions of several cytokine receptors has been identified and proven to be critical for transducing signals (Kouro, et al., 1996). The shared βc subunit is believed to be the major signal transducer and executor for IL-3, IL-5 and GM-CSF. Previous studies reveal that not only the cytoplasmic region (Sakamaki, et al., 1992) but also the extracellular domain (D'Andrea, et al., 1994; D'Andrea, et al., 1996; Jenkins, et al., 1995; Stomski, et al., 1996) are important in modulating the growth signals transduced by βc subunit. A membrane proximal region (between Arg456 and Glu517) has been shown to be essential for activation of tyrosine kinases and critical for mitogenesis (Kinoshita, et al., 1995; Sato, et al., 1993). The distal region (between Leu626 and Ser763) has been shown to be responsible for activation of Ras pathway and for anti-apoptotic ability of cytokine (Kinoshita, et al., 1995; Sato, et al., 1993). Interestingly, IL-5, IL-3 and GM-CSF all activate seemingly identical signals in in vitro cell culture system via the βc subunit, however, in the pluripotent stem cells these three cytokines have overlapping, but distinct specificity on the cell lineage preference when stimulating the differentiation of hematopoietic progenitor cells. The lineage specificity of these receptor complexes is believed to be contributed by the α subunit, however, its exact molecular mechanism remains to be determined.

3. GENERAL CYTOKINE-INDUCED SIGNALING EVENTS

Although the specific role of each individual signaling event in cytokine-activated biological effects is largely unknown, considerable detail about these molecules have been

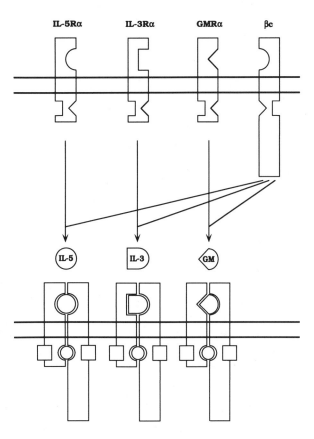

Figure 2. Architecture of hIL-5 receptor complex: Sharing the receptor common β chain in myeloid receptor family. The myeloid receptor family includes receptors of IL-5, IL-3 and GM-CSF. Several common features have been described among members of this family. First, the functional receptor complex contains two subunits, α subunit and β subunit. Second, the α subunit is a low-affinity ligand-binding site and is responsible for ligand specificity. Third, the β subunit is shared by members of this family and forms a high-affinity binding site with each α subunit for cognate cytokine. Fourth, the cytoplasmic domains of both α subunit and β subunit are essential for receptor function.

recently obtained. Extensive tyrosine phosphorylation of intracellular proteins can be triggered by cytokine stimulation and it is the non-receptor protein tyrosine kinases (NRPTK) which associate with and phosphorylate the receptor subunits, and activates subsequent cytoplasmic components. Especially, Janus kinase (Jak) family plays an important role in transducing signals of various cytokine from occupied receptors to the down-stream signaling components (for review see Ihle, et al., 1995; Schindler and Darnell, 1995). Requirement of Jak kinase family for the antiviral activity of IFN-α/β (Velazquez, et al., 1992) and -γ (Sakatsume, et al., 1995; Watling, et al., 1993) has been clearly demonstrated and deficiency of Jak kinases completely abrogates the responses of target cells to IFNs (for review see Darnell, et al., 1994). Jak kinases, in turn, phosphorylate signal transducer and trans-activators (STATs) which are also involved in the signaling pathway of many cytokines (Sadowski, et al., 1993; Shuai, et al., 1993). STATs are essential for the IFN signaling (Leung, et al., 1995) and have been shown to be involved in differentiation signaling of IL-6 (Akira, et al., 1994; Nakajima, et al., 1996) and IL-4 (Kaplan, et al., 1996). STATs may alone, or in collaboration with Ras/Raf signaling pathway, activate the expression of several nuclear oncogenes, like c-myc, c-fos, and c-jun, and establish the antiviral state, differentiation functions, proliferation and anti-apoptosis. Similarly, stimulation of target cells with IL-3 (Mui, et al., 1995), IL-5 (Cornelis, et al., 1995; Kouro, et al., 1996; Mui, et al., 1995), or GM-CSF (Mui, et al., 1995) leads to activation of Jak kinases and tyrosine phosphorylation of other cellular factors which culminate in the induction of nuclear oncogenes.

4. GENERAL PROPERTIES OF CYTOKINE-WITHDRAWAL-INDUCED APOPTOSIS

We characterize general biochemical properties of apoptosis induced by cytokine deprivation in two factor-dependent cell lines in order to further study the molecular mechanisms underlying this process. Human erythroleukemic cell line TF-1 has been isolated from bone marrow of a Japanese erythroleukemia patient which maintains surface expression of the progenitor cell-specific marker, CD34. The proliferation of TF-1 is dependent upon IL-3, GM-CSF or erythropoietin (Kitamura, et al., 1989). Ba/F3 is a murine pro-B cell line established from bone marrow of Balb/C mice and proliferates exclusively in the presence of murine IL-3 (Palacios and Steinmetz, 1985). TF-1 cells begin entering apoptosis within 8 hours following cytokine deprivation and about 30% of them apoptose at 24 hours. Apoptotic TF-1 cells manifest the general characteristics of apoptosis observed in most other cell types, including condensed chromatin, blebbing of cell membrane, fragmentated nuclei and internucleosomal cleavage of genomic DNA (Yen, et al., 1995). The same characteristics are also observed in mIL-3 depleted Ba/F3 cells. Therefore, the combined biochemical and cytochemical analyses indicate that TF-1 and Ba/F3 cells possess a prototype of apoptosis in the absence of their essential cytokines as described in dexamethasome-treated thymocytes. Several features of cytokine-withdrawal-induced apoptosis are described below.

4.1. Cell Cycle Independence

When the kinetics of apoptosis was measured between the asynchronized and synchronized populations, including G1, S and G2/M phases, which were prepared with elutriator, apoptosis appears in all samples within 6 to 8 hours upon cytokine deprivation (unpublished data). These data suggest that the onset of apoptosis in TF-1 cells is cell-cycle stage-independent.

4.2. De Novo Protein Synthesis Dependence

Furthermore, in the presence of 10 to 50 μg/ml of cyclohexamide apoptosis was abrogated in both TF-1 and Ba/F3 cells (unpublished data), indicating that the synthesis of a new protein is essential to the onset of death program in factor-dependent cells.

4.3. Nuclease and Ribonuclease Activation

Extending from the observation of appearance of nucleosomal DNA ladder and specific ribosomal RNA cleavage during some apoptosis, we performed in-gel DNase and

Table 1. Properties of cytokine-withdrawal-induced apoptosis

1.	Appearance of apoptotic bodies, chromatin condensation, and oligonucleosomal ladder
2.	Cell cycle dependent
3.	De novo protein synthesis-dependent
4.	Activation of Ca^{++}/Mg^{++}-DNase
5.	Activation of RNase
6.	Activation of ICE-like protease
7.	Protected by bcl-2 overexpression
8.	Down-regulation of Mcl-1

RNase assays attempting to identify and characterize the corresponding DNase and RNase. A single protein species with the molecular weight of 50 KDa was detected in the apoptotic cell lysate when calcium or magnesium was present (unpublished data). This candidate DNase is similar to the DNase reported by others in apoptotic thymocytes. We also identified three RNase species (p18, p22, and p24) which appeared after 24 hr of cytokine depletion (unpublished data). These RNases function optimally in pH between 7 and 8. The significance of activating RNases during apoptosis of hematopoietic cells remains to be investigated.

4.4. ICE-Related Protease Dependence

Resembling that of many other apoptosis processes, activation of interleukin 1β-converting enzyme (ICE) related protease(s) was found to be essential for cytokine withdrawal-induced apoptosis. In the presence of 300 μg/ml of Z-VAD-FMK, apoptosis of Ba/F3 cells was strongly suppressed within 12 hours (unpublished data). Several candidate substrates for ICE protease cleavage are currently under investigation.

4.5. Bcl-2 and Mcl-1

Although in the factor-dependent cell lines endogenous Bcl-2 levels are easily detectable and do not respond to the onset of apoptosis, exogenous over-expression of Bcl-2 protein in both TF-1 and Ba/F3 cells confers prolonged survival on transfectants after cytokine deprivation (unpublished data). Intriguingly, the endogenous Mcl-1 protein, another member of Bcl-2 family, rapidly degraded prior to the onset of apoptosis, and could be re-expressed after supplemented with human GM-CSF within one hour. Anti-sense oligonucleotide experiments demonstrate that down-regulation of Mcl-1 is sufficient to trigger apoptosis (unpublished data). Our data suggest a pivotal role of Mcl-1 protein, but not the Bcl-2 protein, in the anti-apoptotic function of cytokine in hematopoietic cells.

5. MODEL FOR STUDYING APOPTOSIS-SUPPRESSION BY CYTOKINE

Another way of exploring molecular mechanisms underlying cytokine starvation induced-apoptosis is to tackle the question the other way around and ask how cytokine suppresses apoptosis of its target cells. As was mentioned earlier that numerous molecular events are activated and induced intracellularly, but little is known about their physiological consequences. In a factor-dependent cell line, the essential cytokine usually simultaneously stimulates mitogenesis and suppresses apoptosis. Some molecules may involve in the differentiation program of progenitor cells which is frequently damaged during the establishment of these cell lines. These properties make detail assignment of a biological role to each signaling molecule very difficult. Fortunately, from the beginning of our study of human interleukin 5, the only one available established human IL-5 "responsive" cell line, TF-1, has a "sick" appearance in the presence of hIL-5. It was commonly described as a short term survival effect. The "survival effect" could be a result of either a Bcl-2-like effect of "preservation" of cells in a certain stage of the cell cycle without cell cycle progression, or of equal rates of cell growth and death in a given population. Either mechanism implies that the mitogenic promotion activity and apoptosis suppression activity of a cytokine may be uncoupled under this particular situation. Morphological appearance as well as

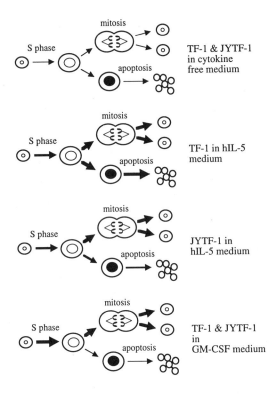

Figure 3. A schematic presentation of the mitogenesis stimulation and apoptosis suppression in TF-1 and JYTF-1 cells by GM-CSF & IL-5. In contrast to that in cytokine-deprivation condition, the mitogenesis (i.e. DNA synthesis activity) of cells in S phase is enhanced by specific cytokines (thicker arrows) above the basal level activity (thin arrow). These cells with newly synthesized DNA will enter either apoptosis or mitosis depending upon supplemented cytokine. In the presence of a specific cytokine (GM-CSF or IL-5), the apoptosis of its dependent cells is suppressed, and the ratio of cells undergoing mitosis to those undergoing apoptosis is greatly enhanced. The extent of apoptosis suppression is not only dependent upon cytokines (GM-CSF vs. IL-5), but also upon the genetic background of the target cells (TF-1 vs. JYTF-1). As indicated, IL-5 is capable of suppressing the ratio of apoptosis / mitosis in JYTF-1 cells, but not in TF-1 cells, and it does not affect the DNA synthesis activity.

biochemical evidence indicate that IL-5 is capable of stimulating mitogenesis of TF-1 cells, but it is unable to suppress apoptosis. Moreover, we have successfully isolated a variant from TF-1 (JYTF-1) wherein death suppression ability of IL-5 is dramatically increased, while its mitogenic promoting ability remains unchanged (Yen, et al., 1995). Interestingly, JYTF-1 cells remain to be GM- or IL-3-dependent and undergo apoptosis upon deprivation of cytokines. Our data indicate that possible target(s) of the genetic alteration in JYTF-1 cells may reside at the very upstream of IL-5 signaling pathway, and that the improvement of apoptosis suppression does not due to the constitutive activation of anti-apoptosis genes, or due to the secretion of secondary autocrine (Yen, et al., 1995). Therefore, we conclude that there is a defect in the IL-5R-mediated apoptosis suppression signaling pathway in TF-1 cells. The apoptosis suppression ability is restored in JYTF-1 cells and this allows the long term growth of JYTF-1in the IL-5 containing medium. The existence of two cell lines whose phenotypes differ in coupling and uncoupling between apoptosis suppression and mitogenesis stimulation opens an avenue to discern apoptosis suppression-specific signaling pathway and mitogenesis promotion signaling pathway.

6. DIFFERENTIAL REGULATION OF MITOGENECITY AND APOPTOSIS SUPPRESSION BY IL5Rα

Further molecular biological analysis of TF-1 and JYTF-1 cells reveals the elevation of mRNA expression of IL5Rα gene in JYTF-1 cells (Yen, et al., 1995). Northern blot analysis and surface antibody staining studies indicate that JYTF-1 express five to eight fold more IL5Rα than TF-1 cells. There is no alteration in the expression level of βc protein, however, the level of tyrosine phosphorylation of βc increases about twenty fold in JYTF-1 after IL-5 stimulation. Secondly, we discovered that the tyrosine phosphorylation of Jak2, STAT5 and ERK2 proteins and induction of Jun-B gene are responsive to IL-5 stimulation in TF-1 cells and are correlated with the mitogenic promotion activity of IL-5 (unpublished data). In contrast to GM-CSF and IL-3, IL-5 does not activate Jak1 and c-myc in TF-1 cells (unpublished data). However, with elevated expression of IL5Rα subunit in JYTF-1 cells, the Jak1/c-myc activation is restored in addition to the increased activation of Jak2/STAT5/ERK2/ jun-B in the presence of IL-5 (unpublished data). Lastly, over-expression of the exogenous IL5Rα in TF-1 cells by retroviral infection not only greatly improved the efficiency of apoptosis suppression of IL-5, but also restored the activation of JAK1 and induction of c-myc (unpublished data). Since these IL5Rα transfectants have the same mitogenic ability as TF-1 and JYTF-1 in IL-5-containing culture medium, we conclude that increaseing of the number of IL5Rα on cell surface specifically suppresses the occurrence of apoptosis in factor-dependent cells. Low level expression of IL5Rα in TF-1 is already sufficient for TF-1 to fully respond to the mitogenic stimulation of IL-5. It is therefore proposed

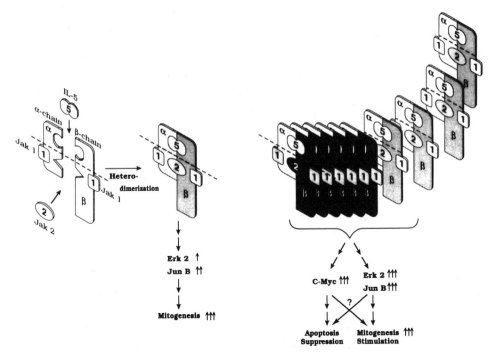

Figure 4. A hypothetical two-stage model for activation of IL-5 receptor functions. In the absence of IL-5, Jak1 is pre-associated with receptor α and β subunit seperately. IL-5 triggers heterodimerization of two receptor subunits and the recruitment and activation of Jak2 (the first stage). At this stage, signaling molecules including Erk2 kinase and JunB oncogene are activated or induced in conjunction with mitogenic stimulation. The tyrosine phosphorylated Jak2 and β subunit are shown in gray. Immediately following the formation of heterodimers, possibly due to conformational alteration, homodimerization of the receptor β subunit occurs through further aggregation of receptor dimers (the second stage). Both Jak2 and β subunit are now further activated and Jak1 becomes slightly phosphorylated in tyrosine. In addition to activation of Erk2 kinase and induction of JunB, c-myc gene is now induced. Phenotypically, not only mitogenesis is stimulated, but also apoptosis is fully suppressed which is likely resulting from Jak1 activation and c-myc induction, or from the concurrent induction of both JunB and c-myc genes. In the presence of large number of receptor subunits, like in JYTF-1, IL-5 receptors are fully activated by IL-5 without interruption between these two stages. In contrast, when α subunit is limited and the number of αβ dimer is low, receptor dimers are scattered around on cell surface and are unable to aggregate. The activation process is forced to stop at the first stage and manifests an un-coupled phenotype of growth stimulation and death prevention, like in TF-1 cells.

that in response to cytokine binding, activation of mitogenicity requires different level of IL5Rα subunit from that requires for anti-apoptosis activity.

7. CONCLUSIONS

At present, the molecular mechanism of cytokine-withdrawal-induced apoptosis remains elusive. However, we have learned many biochemical properties of cytokine withdrawal-induced apoptosis from the factor-dependent cell lines. Further study on the identification of apoptosis-specific protease and on the regulation mechanism of Mcl-1 anti-apoptosis protein by cytokine-deprivation will greatly facilitate our understanding on the nature of death of hematopoietic cells. Alough our data suggest a strong correlation between Jak1/c-myc activation and apoptosis suppression in cytokine-dependent cells, it remains to be investigated about the exact roles of these signaling molecules in the apoptosis suppression. In summary, studying the molecular mechanisms of the growth and death control of cytokine-dependent cells will shed light on the mechanism of leukemogenesis and resume novel approachs toward leukemia therapy.

REFERENCES

Akira, S., Nishio, Y., Inoue, M., Wang, X. J., Wei, S., Matsusaka, T., Yoshida, K., Sudo, T., Naruto, M. and Kishimoto, T. (1994). Molecular cloning of APRF, a novel IFN-stimulated gene factor 3 p91-related transcription factor involved in the gp130-mediated signaling pathway. Cell 77, 63–71.

Cornelis, S., Fache, I., Van-der-Heyden, J., Guisez, Y., Tavernier, J., Devos, R., Fiers, W. and Plaetinck, G. (1995). Characterization of critical residues in the cytoplasmic domain of the human interleukin-5 receptor alpha chain required for growth signal transduction. Eur J Immunol 25, 1857–1864.

D'Andrea, R., Rayner, J., Moretti, P., Lopez, A., Goodall, G. J., Gonda, T. J. and Vadas, M. (1994). A mutation of the common receptor subunit for interleukin-3 (IL-3), granulocyte-macrophage colony-stimulating factor, and IL-5 that leads to ligand independence and tumorigenicity. Blood 83, 2802–2808.

D'Andrea, R. J., Barry, S. C., Moretti, P. A. B., Jones, K., Ellis, S., Vadas, M. A. and Goodall, G. J. (1996). Extracellular truncations of hβc, the common signaling subunit for interleukin-3 (IL-3), granulocyte-macrophage colony-stimulating factor (GM-CSF), and IL-5, lead to ligand-independent activation. Blood 87, 2641–2648.

Darnell, J., Jr., Kerr, I. M. and Stark, G. R. (1994). Jak-STAT pathways and transcriptional activation in response to IFNs and other extracellular signaling proteins. Science 264, 1415–1421.

Ihle, J. N., Witthuhn, B. A., Quelle, F. W., Yamamoto, K. and Silvennoinen, O. (1995). Signaling through the hematopoietic cytokine receptors. Annu Rev Immunol 13, 369–398.

Jenkins, B. J., D'Andrea, R. and Gonda, T. J. (1995). Activating point mutations in the common beta subunit of the human GM-CSF, IL-3 and IL-5 receptors suggest the involvement of beta subunit dimerization and cell type-specific molecules in signalling. EMBO J 14, 4276–4287.

Kaplan, M. H., Schindler, U., Smiley, S. T. and Grusby, M. J. (1996). Stat6 is required for mediating responses to IL-4 and for the development of Th2 cells. Immunity 4, 313–319.

Kinoshita, T., Yokota, T., Arai, K. and Miyajima, A. (1995). Suppression of apoptotic death in hematopoietic cells by signalling through the IL-3/GM-CSF receptors. EMBO J 14, 266–275.

Kishimoto, T., Taga, T. and Akira, S. (1994). Cytokine signal transduction. Cell 76, 253–262.

Kitamura, T., Tange, T., Terasawa, T., Chiba, S., Kuwaki, T., Miyagawa, K., Piao, Y. F., Miyazono, K., Urabe, A. and Takaku, F. (1989). Establishment and characterization of a unique human cell line that proliferates dependently on GM-CSF, IL-3, or erythropoietin. J Cell Physiol 140, 323–334.

Kouro, T., Kikuchi, Y., Kanazawa, H., Hirokawa, K., Harada, N., Shiiba, M., Wakao, H., Takaki, S. and Takatsu, K. (1996). Critical proline residues of the cytoplasmic domain of the IL-5 receptor alpha chain and its function in IL-5-mediated activation of JAK kinase and STAT5. Intl Immunology 8, 237–245.

Leung, S., Qureshi, S. A., Kerr, I. M., Darnell, J., Jr. and Stark, G. R. (1995). Role of STAT2 in the alpha interferon signaling pathway. Mol Cell Biol 15, 1312–1317.

Mui, A. L., Wakao, H., O'Farrell, A. M., Harada, N. and Miyajima, A. (1995). Interleukin-3, granulocyte-macrophage colony stimulating factor and interleukin-5 transduce signals through two STAT5 homologs. EMBO J 14, 1166–1175.

Nakajima, K., Yamanaka, Y., Nakae, K., Kojima, H., Ichiba, M., Kiuchi, N., Kitaoka, T., Fukada, T., Hibi, M. and Hirano, T. (1996). A central role for Stat3 in IL-6-induced regulation of growth and differentiation in M1 leukemia cells. EMBO J 15, 3651–3658.

Palacios, R. and Steinmetz, M. (1985). Il-3-dependent mouse clones that express B-220 surface antigen, contain Ig genes in germ-line configuration, and generate B lymphocytes in vivo. Cell 41, 727–734.

Sadowski, H. B., Shuai, K., Darnell, J., Jr. and Gilman, M. Z. (1993). A common nuclear signal transduction pathway activated by growth factor and cytokine receptors. Science 261, 1739–1744.

Sakamaki, K., Miyajima, I., Kitamura, T. and Miyajima, A. (1992). Critical cytoplasmic domains of the common beta subunit of the human GM-CSF, IL-3 and IL-5 receptors for growth signal transduction and tyrosine phosphorylation. EMBO J 11, 3541–3549.

Sakatsume, M., Igarashi, K., Winestock, K. D., Garotta, G., Larner, A. C. and Finbloom, D. S. (1995). The Jak kinases differentially associate with the alpha and beta (accessory factor) chains of the interferon gamma receptor to form a functional receptor unit capable of activating STAT transcription factors. J Biol Chem 270, 17528–17534.

Sato, N., Sakamaki, K., Terada, N., Arai, K. and Miyajima, A. (1993). Signal transduction by the high-affinity GM-CSF receptor: two distinct cytoplasmic regions of the common beta subunit responsible for different signaling. EMBO J 12, 4181–4189.

Schindler, C. and Darnell, J., Jr. (1995). Transcriptional responses to polypeptide ligands: the JAK-STAT pathway. Annu Rev Biochem 64, 621–651.

Shuai, K., Ziemiecki, A., Wilks, A. F., Harpur, A. G., Sadowski, H. B., Gilman, M. Z. and Darnell, J. E. (1993). Polypeptide signalling to the nucleus through tyrosine phosphorylation of Jak and Stat proteins. Nature 366, 580–583.

Stomski, F. C., Sun, Q., Bagley, C. J., Woodcock, J., Goodall, G., Andrews, R. K., Berndt, M. C. and Lopez, A. F. (1996). Human interleukin-3 (IL-3) induces disulfide-linked IL-3 receptor alpha- and beta-chain hetrodimerization, which is required for receptor activation but not high-affinity binding. Mol Cell Biol 16, 3035–3046.

Takaki, S., Tominaga, A., Hitoshi, Y., Mita, S., Sonoda, E., Yamaguchi, N. and Takatsu, K. (1990). Molecular cloning and expression of the murine interleukin-5 receptor. EMBO J 9, 4367–4374.

Velazquez, L., Fellous, M., Stark, G. R. and Pellegrini, S. (1992). A protein tyrosine kinase in the interferon alpha/beta signaling pathway. Cell 70, 313–322.

Watling, D., Guschin, D., Muller, M., Silvennoinen, O., Witthuhn, B. A., Quelle, F. W., Rogers, N. C., Schindler, C., Stark, G. R. and Ihle, J. N., et al. (1993). Complementation by the protein tyrosine kinase JAK2 of a mutant cell line defective in the interferon-gamma signal transduction pathway. Nature 366, 166–170.

Yen, J. J., Hsieh, Y. C., Yen, C. L., Chang, C. C., Lin, S. and Yang-Yen, H. F. (1995). Restoring the apoptosis suppression response to IL-5 confers on erythroleukemic cells a phenotype of IL-5-dependent growth. J Immunol 154, 2144–2152.

14

GENETIC REGULATION OF APOPTOSIS IN THE MOUSE THYMUS

Barbara A. Osborne,[1,2] Sallie W. Smith,[1] Kelly A. McLaughlin,[2] Lisa Grimm,[2] Grant Morgan,[1] Rebecca Lawlor,[1] and Richard A. Goldsby[3]

[1]Department of Veterinary and Animal Sciences
[2]Program in Molecular and Cellular Biology
University of Massachusetts
Amherst, Massachusetts 01003
[3]Department of Biology Amherst College
Amherst, Massachusetts 01002

1. INTRODUCTION

The role of apoptosis in the maintenance of homeostasis in cells of the immune system recently has become apparent [1,2]. Indeed some of the best characterized examples of apoptosis are found in the immune system. For example, apoptosis is the mechanism used during negative selection in the thymus to remove self-reactive T cells[3–5]. Thymic T cells also are quite susceptible to induction of apoptosis by either glucocorticoids or ionizing radiation[6,7]. Additionally, more recent data indicate that peripheral lymphocytes undergo apoptosis following a variety of different stimuli and, in many instances, cell death in activated peripheral T and B cells may be traced to Fas/FasL interactions[8–11]. Peripheral T cells also have been shown to susceptible to induction of cell death by TNFα[12].

1.1. Cell Death in the Thymus

The first clear indication that apoptosis may be physiologically relevant to thymocytes, comes from the early work of Wyllie who showed that injection of glucocorticoid induced massive apoptosis in the thymus[6]. While these experiments involved the introduction of synthetic glucocorticoids into animals, more recent evidence from Ashwell and colleagues demonstrate that naturally occurring glucocorticoids induce apoptosis *in vivo*[13,14]. Another stimulus known to induce apoptosis in the thymus is ionizing radiation[7].

It has been known for many years that immature thymocytes are susceptible to death and data from Kappler and Marrack demonstrate that superantigens induce death in a large proportion of thymocytes carrying the particular Vβ molecule that binds the given superantigen[15]. Other investigators, using TCR transgenic mice, have shown that encounter with the cognate antigen in the thymus induces deletion of the immature CD4[+],CD8[+] thymocytes[16–18].

The conclusions that negative selection involves the induction of apoptosis in this immature population of thymocytes has been extended using thymic organ culture[3,19,20].

2. GENES THAT MEDIATE APOPTOSIS

There are many indications from several independent systems that new gene synthesis is required for the induction of some forms of cell death[21]. In particular, we and others have found that cell death in T cells requires transcription[22–24]. In an attempt to elucidate the molecules responsible for the initiation of apoptosis in T cells, we created a cDNA library from mRNA isolated from dying thymocytes. By subtractive screening strategies, we have isolated several genes whose expression is either up-regulated or repressed during apoptosis. We focused our search on genes expressed during cell death induced by negative selection. In order to closely mimic the conditions of negative selection, we chose to use a T cell receptor (TCR) transgenic mouse previously demonstrated by Murphy and colleagues to be susceptible to the induction of apoptosis by intraperitoneal injection of antigen[5]. In this system, mice are injected with the ovalbumin peptide OVA 323–336 and initiation of apoptosis proceeds within a few hours following injection of peptide. We used mRNA isolated from the thymus of such animals 3, 4 and 5 hours following injection of antigen. This mRNA was used to construct a cDNA library that was screened using a plus/minus screening strategy.

The method chosen for screening allowed the isolation of several genes whose expression pattern is up-regulated following cell death, as well as two genes, *apt-1* and *apt-3*, that are repressed during apoptosis. In addition to repressed genes, we found several genes induced during apoptosis in T cells. One gene, originally called *apt-2* but later identified as the immediate-early gene *nur77*, is induced within the first 45 minutes following crosslinking of the TCR This gene is expressed at high levels following induction of cell death and is expressed throughout the process of death. *Nur77* is a member of the steroid-thyroid hormone receptor superfamily, a gene family characterized by its activity as a transcription factor. The expression of *nur77* was first observed in serum stimulated 3T3 cells[25]. Subsequently, others have found the expression of *nur77* in PC12 cells induced to differentiate into neuronal cells[26]. Therefore the expression of this gene has been associated with proliferation, differentiation and, as a consequence of our observations, with apoptosis. This is less surprising in light of data from Green and co-workers who have shown that up-regulation of *c-myc*, a gene traditionally associated with proliferation, is required for TCR-mediated death in T cell hybridomas[23].

The induction of *nur77* occurs shortly following crosslinking of the TCR and is inhibited by actinomycin D, a transcriptional inhibitor that also inhibits cell death in DO11.10 cells. Up-regulation of *nur77* also is seen in the thymus of animals treated with peptide to induce death and in thymic organ cultures treated with peptide. Interestingly, treatment of animals or the DO11.10 cell line with glucocorticoids such as dexamethasone, induces cell death but does not result in up-regulation of *nur77*. These data indicate to us that the expression of *nur77* is specific to signals through the TCR.

To determine if the expression of *nur77* is essential for cell death, we made constructs that express the gene in either a truncated sense or antisense orientation. When DO11.10 cells are transfected with a truncated sense construct in conjunction with CD20, a reporter gene used to mark the transfected cells, and induced to die, only 2% CD20+ cells are observed in the live population of cells, indicating that the truncated sense construct does not protect against apoptosis. However, when the antisense construct is used, a significant number (12.9%) of CD20+ cells are observed in the live population of cells. These data pro-

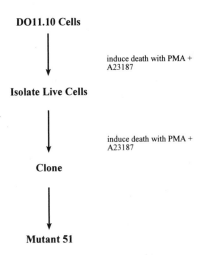

Figure 1. Derivation of mutant 51.

vided convincing evidence, on a cell-by cell basis, that inhibition of *nur77* by antisense prevents cell death[27]. Thus *nur77* is required for cell death in T cell hybrids. Data from Woronicz et al.[28] indicating a *nur77* dominant negative mutant also acts to inhibit TCR-mediated death in T cell hybrids strengthens and extends our conclusions. Most recently, Calnan et al. and Weih et al. have provided convincing evidence that lines of transgenic mice carrying a dominant negative *nur77* targeted to the thymus, have defects in negative selection providing further evidence for a requirement of *nur77* function in thymic cell death[29,30].

3. MUTANT CELL LINES AS A TOOL TO STUDY APOPTOSIS

In order to further our understanding of the events that mediate cell death in T cells, we created a series of mutant cell lines from the parental line, DO11.10, by selection for cells that survive PMA + A23187 induced death (Figure 1). One such mutant, mutant 51 was isolated and this line was shown to be resistant to the induction of cell death via TCR crosslinking or by PMA +A23187. Mutant 51 however is susceptible to death induced via ionizing radiation or dexamethasone (Table 1). While mutant 51 does produce *nur77*

Table 1. Comparison of DO11.10 versus mutant 51

	DO11.10 Clone 2	Mutant 51
Ability to die in response to PMA + A23187	+	−
Ability to die in response to TCR crosslinking	+	−
Ability to die in response to dexamethasone	+	+
Ability to die in response to γ-irradiation	+	+
IL-2 production following TCR crosslinking	+	−
Induction of *nur77* mRNA following TCR crosslinking	+	+
Phosphorylation of Nur77 protein following TCR crosslinking	+	−
Release of intracellular [Ca++] following TCR crosslinking	+	+/−

mRNA in response to TCR signals, very little Nur77 protein is made and the protein is not phosphorylated. In normal situations, TCR signaling induces *nur77* mRNA and Nur 77 protein that is heavily phosphorylated. Whether the lack of phosphorylated Nur77 protein in mutant 51 is directly associated with the inability of this cell line to die in response to TCR signals remains to be determined.

4. THE ROLE OF OXYGEN IN T CELL DEATH

Despite the differences in the pathways by which different stimuli initiate the apoptosis program in thymocytes, there are similarities in their responses to pharmacological agents that affect the levels of reactive oxygen intermediates (ROI). Treatment of thymocytes with the thiol, N-(2-mercaptoethyl)-1,3 propanediamine (WR-1065), blocks the induction of death in thymocytes by gamma radiation, glucocorticoids or treatment with calcium ionophores[31]. A number of other compounds with a potential for antioxidant activity, including N-acetyl cysteine[32], retinoic acid[33], and vitamin E[34], have been reported to display anti-apoptotic activity in some systems. Also, it has been observed that increasing the level of enzymes that protect against oxidative stress inhibits the induction of apoptosis. Specifically, resistance to growth factor withdrawal-induced apoptosis was conferred on an IL-3 dependent cell line by transfection with a glutathione peroxidase expressing construct[35].

The many observations of the protective effects of ROI scavengers on thymocyte apoptosis induced by signals from the TCR (or its pharmacological equivalents), glucocorticoids or γ-radiation, suggested that this might be a common feature shared by these otherwise different inductive pathways. Despite the correlation between ROI scavenging and inhibition of apoptosis, it was not clear how the pharmacological effects of antioxidants on these pathways might to be interpreted. Although it seemed likely that their action could be traced to a direct lowering of the level of ROIs, other mechanisms could not be ruled out. We initiated a series of experiments to address the question of whether the induction of any or all these T cell deaths are dependent upon oxygen derived species[36]. This was done directly by determining whether or not the capacity of a given agent to induce apoptosis is affected by the withdrawal of molecular oxygen. If T cell apoptosis induced by glucocorticoids or TCR-like signals is unaffected by anaerobiosis, it must be concluded that ROIs do not have an essential role in the induction of death by these stimuli.

4.1. The Induction of Death in Thymocytes Requires Oxygen

To examine the role of oxygen in apoptosis in thymocytes, an anaerobic chamber was constructed and thymocytes were induced to die via various stimuli in the absence of oxygen. To determine the extent to which thymocytes respond to death-inducing stimuli under aerobic and anaerobic conditions, thymocytes were maintained under an air/CO_2 atmosphere or in N_2/CO_2 during treatment. Figure 2 shows that treatment of thymocytes with concentrations of apoptosis-inducing concentrations of PMA/A23187 produces approximately 60% death in air/CO_2 but yields only 1% under a N_2/CO_2 atmosphere. Figure 2 also shows that treatment of thymocytes with dexamethasone induced approximately 65% death under air/CO_2. However, the same dose produced only 8% death in these cells when administered under a N_2/CO_2 atmosphere. Killing by PMA/A23187 and dexamethasone show similar oxygen dependencies in both T cell hybridoma and thymocytes.

Thymocytes stimulated to die through the addition of phorbol ester plus calcium ionophore were also protected when 20 mM n-acetyl- cysteine (NAC), a potent antioxidant, was

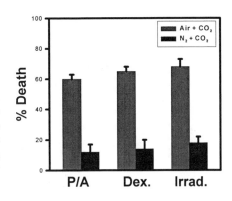

Figure 2. Induction of death by gamma-irradiation, PMA/A23187, or dexamethasone in the presence or absence of molecular oxygen. Treatment of thymocytes with either 10 nM phorbol myristate (PMA) and 250 nM calcium ionophore A23187, 2.5 μM dexamethasone or 100 cGys of γ-irradiation. Cells were maintained in either Air/CO_2 or N_2/CO_2 as indicated for 14–18 hours before FACScan analysis. Results are expressed as a ±SD of triplicate determination.

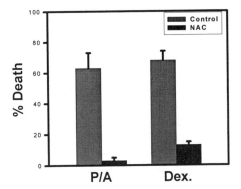

Figure 3. Effects of N-acetylcysteine (NAC) on DO11.10 cell death induced by PMA/A23187 or dexamethasone. Cells were treated with 10 nM PMA plus 250 nM A23187, or 2.5 μM dexamethasone in the presence or absence of 60 mM NAC, incubated under standard conditions and analyzed for induction of apoptosis. Results are expressed as a ±SD of triplicate determination.

added to cultures prior to stimulation. NAC also protected thymocytes from dexamethasone and radiation-induced apoptosis (Figure 3). These data provide compelling evidence for a requirement for oxygen during the induction of many apoptotic pathways in thymocytes.

5. THE 26S PROTEASOME IS A REQUIRED COMPONENT OF CELL DEATH IN THE THYMUS

Quite recently, we have found that the function of the 26S proteasome in required in many forms of thymocyte death. The 26S proteasome is a conserved multicatalytic proteolytic complex present in all eukaryotic cells[37] that is responsible for the degradation of most cellular proteins[38]. It is composed of a 20S catalytic core and associated regulatory proteins. This complex plays a critical role in the ubiquitin-proteasome-dependent proteolytic pathway, where it catalyzes the rapid degradation of proteins covalently linked to chains of ubiquitin. Because of the involvement of proteolytic events in cell death and the known function of the proteasome in proteolytic processes, we examined whether the proteasome also played a central role in cell death. Our recent data demonstrate a requirement for the proteasome in apoptosis initiated in primary thymocytes by diverse stimuli[39].

To determine if the proteasome plays a role in apoptosis, compounds known to inhibit proteasome function were added to primary thymocytes prior to or simultaneously upon the induction of cell death by either ionizing radiation, glucocorticoids or the phorbol ester, PMA, which mimics T cell receptor engagement in thymocytes. Table 2 provides evidence that LLnL and LLM significantly inhibit apoptosis in thymocytes. Because

Table 2. Proteasome function is required for apoptosis in thymocytes

| | % Inhibition of apoptosis* | | | |
| | Treatment | | | |
Inhibitor	Control	Dexamethasone	γ-Radiation	PMA
LLM(50μM)	−	55	50	55
LLnL(50μM)	−	90	78	85
MG132(10μM)	−	90	97	70
Lactacystin(10μM)	−	65	80	58
E64(50μg/ml)	5	8	5	10

*Cells were analyzed 8 hr after the induction of apoptosis in the presence of the indicated proteasome inhibitor. Eight hours after induction of death, thymocytes were analyzed for FITC or Yo-Pro-3 iodide staining on the FACScan. MTT assays were also used to measure viability. The % inhibition of death was calculated as follows: 100 - [% dead cells in inhibitor-treated population/ % dead cells in control population x 100].

these peptides can inhibit calpain as well as the proteasome[38], we included in these experiments E64-ester, a cell-permeable inhibitor of the cysteine proteases calpain I and II, and the lysosomal protease, cathepsins B, H and L. None of the concentrations of E64 tested blocked cell death in any of the conditions tested, further confirming the specificity of the proteasome inhibitors.

Lactacystin was shown to bind irreversibly and covalently to the N-terminal threonine of one of the β subunits of the proteasome[40]. Lactacystin has not been found to block the activity of other proteases including calpain, cathepsin B, chymotrypsin, trypsin and papain[40]. Thus lactacystin is a highly specific and irreversible inhibitor of the proteasome *in vitro* and in intact cells. In thymocytes, this compound was able to block apoptosis induced by radiation, dexamethasone, or PMA (Table 2) demonstrating that the proteasome specifically is required for apoptosis in these cells.

One important enzyme that is proteolytically inactivated in apoptotic cells is the DNA repair enzyme PARP[41]. To test whether inhibitors of proteasome activity also affect the apoptotic process before or after PARP cleavage, primary thymocytes were induced to die by either radiation or dexamethasone in the presence of proteasome inhibitors and PARP cleavage assayed by western blot analysis. We found that in unstimulated cells, intact 116 kDa PARP is detected but after exposure to either radiation or dexamethasone, PARP was cleaved to an 85 kDa polypeptide. This cleavage was inhibited by proteasome inhibitors.

In some cell culture systems, proteasome inhibitors have been implicated in the induction of apoptosis[42]. This is not surprising in light of the evidence for the requirement of the proteasome in cell cycle progression[43–44]. In an actively dividing cell population, inhibition of cell cycle progression might well be expected to be toxic in some instances. However, we have noted that prolonged exposure of thymocytes to these inhibitors over time increases the amount of background death, suggesting that prolonged inhibition of proteasome function might also be toxic to these cells (Figure 4). As shown in Figure 4, when DO11.10 cells are treated with proteasome inhibitors, cell death proceeds in the absence of inductive strategies.

It is not yet clear what precise role the proteasome performs in cell death. It is possible it cleaves a particular product that initiates a proteolytic cascade, for example, the proteolytic processing of a proICE-like family member into an active protease. Because CPP32, a well characterized ICE family member cleaves PARP[45], our data place the action of the proteasome before the activation of CPP32. Another possible scenario is that the pro-

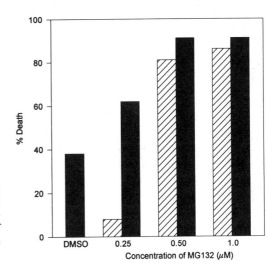

Figure 4. Proteasome inhibitors induce death in DO11.10. DO11.10 cells were treated with MG132 at the indicated concentrations either in the presence of dexamethasone (grey bar) or absence of apoptosis-inducing agent (hatched bar) and % death assayed 18 hours after treatment.

teasome pathway degrades an inhibitor of apoptosis normally present to keep the cell death pathway in check (just as the onset of the inflammatory response is normally inhibited by IκB). The control and specificity of such a system could reside in the ability of the ubiquitin system to ubiquitinate a particular substrate, targeting it for destruction (Figure 5).

6. THE ROLE OF p53 IN THYMOCYTE APOPTOSIS

In a collaboration with Tyler Jacks, we and others also have described a role for p53 in apoptosis in T cells[46,47]. Data from Yonish-Rouach and colleagues suggested that p53 may play a role in the induction of apoptosis[48]. These investigators showed that a mutant form of p53, expressed in a temperature dependent fashion, when transfected into the M1

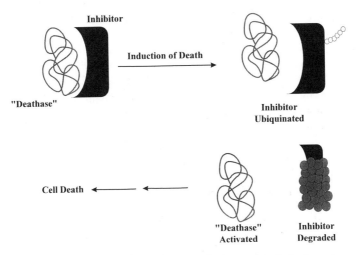

Figure 5. Possible mechanism of proteasome function in apoptosis. A potential role for the proteasome may be the control of degradation of an inhibitor of an essential cell death gene. In such a scenario, the induction of apoptosis would signal the ubiquitination and subsequent degradation of the putative inhibitor, releasing the function of the death gene product.

myeloid leukemia cell line, induced apoptosis at the permissive temperature. Cells grown at the non-permissive temperature did not undergo apoptosis and did not express p53, suggesting that p53 may function as a cell death gene. To test the role of p53 in normal, non-transformed cells, we took advantage of the availability of a p53 null mouse[49]. Using wt +/+, heterozygous +/− and mutant −/− mice, we asked whether p53 is required for apoptosis in thymocytes. The answer was quite clear; in −/− mice, p53 is required for radiation-induced apoptosis but not for glucocorticoid nor TCR-meditated apoptosis. These data suggested that cell death induced by radiation differs from that induced by either TCR or glucocorticoid. In fact, from data described above, we already knew that *nur77* is expressed in TCR-mediated but not glucocorticoid induced death. Taken together, these data indicate there are at least 3 pathways that thymocytes may employ to undergo apoptosis; one pathway induced by gamma radiation and requires p53, another is induced by glucocorticoids and a third, induced by TCR ligation, involves *nur77, egr-1* and *apt-4*. Furthermore, *nur77* is required for the latter pathway.

7. CONCLUSIONS

The decision to die is an active process in many cells. In particular, T lymphocytes appear to require new gene synthesis to die. What signals initiate this process and what genes are required? Some of the signals include T cell receptor engagement, glucocorticoids and ionizing radiation. While the details of glucocorticoid induced apoptosis are not well-understood, it is known that the glucocorticoid receptor (GR) is required and deletion of the DNA-binding domain of the GR interferes with the induction of death[50]. These data imply that GR acts by initiating transcription of genes that are required for the subsequent death of the cells. The identity of these molecules, as well as their mechanism of action, have yet to be elucidated. While Cohen and colleagues have isolated genes induced during glucocorticoid death, their role in this process has not been determined[51].

The mechanism of radiation-induced death requires p53 as described above. The p53 protein can act as a transcription factor suggesting that its role in radiation-induced apoptosis is the initiation of transcription of down-stream gene(s). In fact, a candidate down-stream gene is the recently described p21 gene. The protein encoded by p21 forms a multimeric complex with PCNA resulting in cell cycle arrest[52]. Recent data from the Karin lab indicates that p53 can act to initiate apoptosis in immortalized pituitary cells in the absence of both RNA and protein synthesis[53]. But Attardi et al. recently have shown that in mouse embryo fibroblasts, transcriptional activation is a required function of the p53 protein in the induction of cell death[54]. Thus the role of transcriptional activation by p53 during apoptosis is somewhat controversial at present. How this relates to p53-mediated death in thymocytes is unknown at present.

Cell death initiated through TCR interactions requires the expression of *c-myc* and *nur77*. Data from Green and co-workers, as well as our own data, demonstrate convincingly that these two genes are required for TCR-mediated death[27–28,55]. As well, cell death through this pathway involves the action of *apt-4* a gene isolated from our original cell death library of induced genes. Preliminary data from our lab suggest this gene is critical, although definitive experiments precisely defining its role in apoptosis are not yet available. As seen in other pathways described above, death through TCR interaction requires the expression of genes encoding transcription factors, indicating a role for genes down-stream of *c-myc* and *nur77*. While these genes appear to be induced by specific apoptotic stimuli, it appears as though all apoptotic pathways in thymocytes and most in T cell lines require molecular oxygen. Sup-

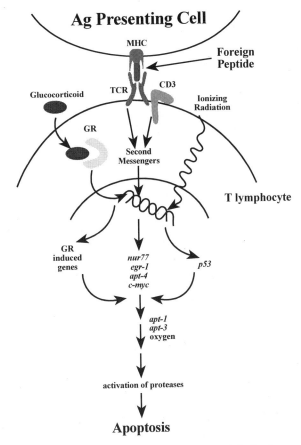

Figure 6. Multiple pathways lead to apoptosis in T cells.

porting this conclusion is the observation that an antioxidant, such as NAC, blocks these cell deaths. Lastly, the function of the proteasome, a multicatalytic complex of proteases is required for many forms of cell death in thymocytes. The precise role of the proteasome as well as the place it occupies in an apoptotic cascade is not yet known but we feel the requirement for the proteasome most probably lies immediately upstream of the activation of the ICE family of proteases but downstream of the contribution of oxygen (see Figure 6).

In conclusion, we now have a better, albeit incomplete, understanding about the events that lead to apoptosis in T cells. We know that genes such as *c-myc* and *nur77* are induced by signals through the TCR, that these genes are required and the expression of these genes is unique to TCR-mediated signals. On the other hand, p53 is required for radiation-induced death but does not appear to play a role in either glucocorticoid or TCR-mediated death. These data indicate the existence of multiple signaling pathways leading to apoptosis in T cells. We propose that these pathways converge and form a final common pathway that results in the ultimate demise of the cell (Figure 6). Our data support a critical role for oxygen and proteasome function in this more common portion of the cell death pathway. Additionally, our data support the notion that the proteasome functions upstream biochemically of the activation of ICE family proteases. To better understand how T cells die by apoptosis, it will be necessary to design experimental protocols that allow us to identify and ultimately dissect the common components of this cell death pathway.

REFERENCES

1. Ellis, R., J. Yuan, and H. R. Horvitz, Mechanisms and functions of cell death. *Annu. Rev. Cell Biol.* 7:663 (1991).
2. Schwartz, L. M. and B. A. Osborne, Programmed cell death, apoptosis and killer genes. *Immunol. Today* 14:582 (1993).
3. Swat, W., L. Ignatowicz, H. von Boehmer, and P. Kisielow, Clonal deletion of immature CD4$^+$ CD8$^+$ thymocytes in suspension culture by extrathymic antigen-presenting cells. *Nature* 351:150 (1991).
4. Smith, C. A., G. Williams, R. Kingston, E. J. Jenkinson, and J. T. Owen, Antibodies to CD3/T-cell receptor complex induces death by apoptosis in immature T cells in thymic culture. *Nature* 337:181 (1989).
5. Murphy, K. M., A. B. Heimberger, and D. Y. Loh, Induction by antigen of intrathymic apoptosis of CD4$^+$ CD8$^+$ TCR^{lo} thymocytes *in vivo*. *Science* 250:1720 (1990).
6. Wyllie, A. H., Glucocorticoid induced thymocyte apoptosis is associated with endogenous endonuclease activation. *Nature* 284:555 (1980).
7. Sellins, K. and J.J. Cohen, Gene induction by γ-irradiation leads to DNA fragmentation in lymphocytes. *J. Immunol.* 139:3199 (1987).
8. Dhein, J., H. Walczak, C. Baumler, K-M. Debatin, and P. H. Krammer, Autocrine T-cell suicide mediated by APO-1/(Fas/CD95). *Nature* 373:438 (1995).
9. Brunner, T., R. J. Mogil, D. LaFace, N.J. Yoo, A. Mahoubi, F. Echeverri, S.J. Martin, W.R. Force, D. H. Lynch, C.F. Ware and D.R. Green, Cell-autonomous Fas (CD95)/Fas-ligand interaction mediates activation-induced apoptosis in T-cell hybridomas. *Nature* 373:441 (1995).
10. Ju, S-T, D. J. Panka, H. Cul, R. Ettinger, M. El-Khatib, D. H. Sherr, B. Z. Stanger, and A. Marshak-Rothstein, Fas (CD95)/FasL interactions required for programmed cell death after T-cell activation. *Nature* 373:444 (1995).
11. Rothstein TL, Wang JKM, Panka, DJ, Foote LC, Wang Z, Stanger B, Cul H, Ju S-T, Marshak-Rothstein A: Protection against Fas-dependent Th1-mediated apoptosis by antigen receptor engagement in B cells. *Nature* 374:163 (1995).
12. Zheng L, Fisher G, Miller RE, Peschon J, Lynch DH, Lenardo MJ: Induction of apoptosis in mature T cells by tumor necrosis factor. *Nature* 377:348 (1995).
13. Vacchio, M., V. Papadopoulus, and J. D. Ashwell, Steroid production in the thymus: implications for thymocyte selection. *J. Exp. Med.* 179:1835 (1994) .
14. King, L.B., M.S. Vacchio, R. Hunziker, D. H. Margulies, and J. D. Ashwell, A targeted glucocorticoid receptor antisense transgene increases thymocyte apoptosis and alters thymocyte development. *Immunity* 5:647 (1995).
15. White, J., A. Herman, A. M. Pullen, R. Kubo, J. Kappler, and P. Marrack, The Vβ-specific superantigen staphylococcal enterotoxin B: Stimulation of mature T cells and clonal deletion in neonatal mice. *Cell* 56:27 (1989).
16. Kisielow, P., H. Bluthmann, U. D. Staerz, M. Steinmetz, and H. von Boehmer, Tolerance in T-cell-receptor transgenic mice involves deletion of nonmature CD4$^+$, 8$^+$ thymocytes. *Nature* 333:742 (1988).
17. Sha, W., C. Nelson, R. Newberry, D. Kranz, J. Russell, and D. Y. Loh, Positive and negative selection of an antigen receptor on T cells in transgenic mice. *Nature* 336:73 (1988).
18. Berg, L., G. Frank and M.M. Davis, The effects of MHC gene dosage and allelic variation on T cell receptor selection. *Cell* 60:1043 (1990).
19. MacDonald, H. R. and R. K. Lees, Programmed death of autoreactive thymocytes. *Nature* 343:624 (1990).
20. Winslow, G. M., M. T. Scherer, J. W. Kappler, and P. Marrack, Detection and biochemical characterization of the mouse mammary tumor virus 7 superantigen (Mls-1a). *Cell* 71:719 (1992).
21. Schwartz, L. M., The role of cell death genes during development. *BioEssays* 13:389 (1991).
22. Cohen, J. J. and R. Duke, Glucocorticoid activation of a calcium-dependent endonuclease in thymocyte nuclei leads to cell death. *J. Immunol.* 132:38 (1984).
23. Shi, Y., M. Szaly, L. Paskar, M. Boyer, B. Singh, and D. R. Green, Activation-induced cell death in T cell hybridomas is due to apoptosis. *J. Immunol.* 144:3326 (1990).
24. Osborne, B. A., S.W. Smith, Z.-G. Liu, K. McLaughlin, L. Grimm and L. M. Schwartz. 1994.Identification of genes induced during apoptosis in T lymphocytes. *Immunological Reviews*, Vol. 141, 301–320.
25. Lau, L. and D. Nathans, Expression of a set of growth-related immediate early genes in BALB/c 3T3 cells: Coordinate regulation with c-fos or c-myc. *Proc. Natl. Acad. Sci. (USA)* 84:1182 (1987).
26. Watson, M. A. and J. Milbrandt, The NGFI-B gene, a transcriptionally inducible member of the steroid receptor gene superfamily: genomic structure and expression in rat brain after seizure induction. *Mol. Cell. Biol.* 9:4213 (1989).

27. Liu, Z.-G., S. W. Smith, K. A. McLauglin, L. M. Schwartz, and B. A. Osborne, Apoptotic signals through the T-cell receptor of a T-cell hybrid require the immediate-early gene *nur77*. *Nature* 36:281 (1994).

28. Woronicz, J. D., B. Calnan, V. Ngo, and A. Winoto, Requirement for the orphan steroid receptor Nur77 in apoptosis of T-cell hybridomas. *Nature* 367:277 (1994).

29. Calnan, B.J., S. Szychowski, F. K-M. Chan, D. Cado, and A. Winoto, A role for the orphan steroid receptor Nur77 in apoptosis accompanying antigen-induced negative selection. *Immunity* 3:273 (1995).

30. Weih, F., Ryseck, R.P., Chen, L., and R. Bravo. Apoptosis of nur77/N10-transgenic thymocytes involves the Fas/Fas ligand pathway. *Proc. Natl. Acad. Sci.* 93:5533 (1996).

31. Ramakrishnan, N. and G. Catravas, N-(2-mercaptoethyl)-1,3-propanediamine (WR-1065) protects thymocytes from programmed cell death. *J. Immunol.* 148:1817 (1992).

32. Mayer, M. and M. Nobel, N-acetyl-L-cysteine is a pluripotent protector against cell death and enhancer of trophic factor-mediated cell survival in vitro. *Proc. Natl. Acad. Sci. USA* 91:7496 (1994).

33. Iwata, M., M. Mukai, Y. Nakai, and R. Iseki, Retinoic acid inhibits activation-induced apoptosis in T cell-hybridoms and thymocytes. *J. Immunol.* 149:3302 (1992).

34. Sandstrom, P. A., M. Mannie, and T. M. Buttke, Inhibition of activation-induced death in T cell hybridomas by thiol antioxidative stress as a mediator of apoptosis. *J. Leukoc. Biol.* 55:221 (1994).

35. Hockenbery, D., Z. Oltvai, X. Yin, C. Milliman, and S. Korsmeyer, Bcl-2 functions in an antioxidant pathway to prevent apoptosis. *Cell* 75:241 (1993).

36. McLaughlin, K. A., Osborne, B.A. and R. A. Goldsby. The role of oxygen in thymocyte apoptosis. *Eur. J. Immunol.*, 26:1170 (1996).

37. Coux, O., Tanaka, K and Goldberg, A.L. Structure and functions of the 20S and 26S proteasomes. *Ann. Rev. Biochem.*, in press (1996).

38. Rock, K.L, Gramm, C., Rothstein, L., Clark, K., Stein, R., Dick, L., Hwang, D. and Goldberg, A.L. Inhibitors of the proteasome block the degradation of most cell proteins and the generation of peptides presented on MHC class I molecules. *Cell* 78:761 (1994).

39. Grimm, L. A., Goldberg, A. L., Poirier, G.G., Schwartz, L.M. and B.A. Osborne. Proteasomes play an essential role in thymocyte apoptosis. *EMBO J.* 15:3835 (1996).

40. Fenteany, G., Standaert R.F., Lane, W.S., Choi, S., Corey E.J., Schreiber, S.L. Inhibition of proteasome activities and subunit-specific amino-terminal threonine modification by lactacystin. *Science* 269:726 (1995).

41. Lazebnik, Y.A., Kaufmann, S.H., Desnoyers, S., Poirier, G.G. and Earnshaw, W.C. Cleavage of poly(ADP-ribose)polymerase by a proteinase with properties like ICE. *Nature* 371:346 (1994).

42. Imajohomi, S., Kawaguchi, T., Sugiyama, S., Tanaka, K., Omura, S. And Kikuchi, H. Lactacystin, a specific inhibitor of the proteasome, induces apoptosis in human monoblast U937 cells. *Biochem. & Biophys. Res. Comm.* 217:1070 (1995).

43. Glotzer, M. Murray, A. and Kirschner, M. Cyclin is degraded by the ubiquitin pathway. *Nature* 349:132 (1991).

44. Pagano, M., Tam, S.W., Theodoras, A.M., Beer-Romero, Del Sal, G., Chau, V., Yew, P.R., Draetta, G.F. and Rolfe, M. Role of the ubiquitin-proteasome pathway in regulating abundance of the cyclin-dependent kinase inhibitor p27. *Science* 269:682 (1995).

45. Nicholson, D.W., Ali, A., Thornberry, N., Vailancourt, J.P., Ding, C.K., Gallant, M., Gareau, Y., Griffin, P.R., Labelle, M., Lazebnik, Y.A., Munday, N.A., Raju, S. M., Smulson, M.E., Yamin, T.-T., Yu, V. and Miller, D.K. Identification and inhibition of the ICE/CED-3 protease necessary for mammalian apoptosis. *Nature* 376:37 (1995).

46. Lowe, S, E. M. Schmitt, S. W. Smith, B. A. Osborne, and T. Jacks, p53 is required for radiation-induced apoptosis in mouse thymocytes. *Nature* 362:847 (1993).

47. Clarke, A.R., C.A. Purdie, D.J. Harrison, R.G. Morris, C.C. Bird, M.L. Hooper, and A. Wyllie, Thymocyte apoptosis induced by p53- dependent pathways. *Nature* 362:849 (1993).

48. Yonish-Rouach, E., D. Resnitzky, J. Lotem, L. Sachs, A. Kimchi, and M. Oren, Wild-type p53 induces apoptosis of myeloid leukemic cells that is inhibited by interleukin-6. *Nature* 352:345 (1991).

49. Jacks, T., L. Remington, B. O. Williams, E. Schmidt, S. Halacimi, R. T. Bronson, and R. A. Weinberg, Tumor spectrum analysis in p53 mutant mice. *Curr. Biol.* 4:1 (1994) .

50. Dieken, E. S. and R. L.Miesfeld, Transcriptional transactivation functions localized to the glucocorticoid receptor N terminus are necessary for steroid induction of lymphocyte apoptosis. *Mol. Cell. Biol.* 92:589 (1992).

51. Owens, G. P., W. Hahn, and J. J. Cohen, Identification of mRNAs associated with programmed cell death in immature thymocytes. *Mol. Cell. Biol.* 11:4177 (1991).

52. Waga, S., G. J. Hannon, D. Beach, and B. Stillman, The p21 inhibitor of cyclin-dependent kinases controls DNA replication by interaction with PCNA. *Nature* 369:520 (1994).

53. Caelles, C., A. Helmberg, and M. Karin, p53-dependent apoptosis in the absence of transcriptional activation of p53 target genes. *Nature* 370:220 (1994).

54. Attardi, L.D., Lowe, S.W., Brugarolas, J., and T. Jacks. Transcriptional activation by p53, but not induction of the p21 gene, is essential for oncogene-mediated apoptosis. *EMBO J.* 15:3693 (1996).

55. Shi, Y., J. M. Glynn, L. J. Guilbert, T. G. Cotter, R. Bissonette, and D. R. Green, Role for c-myc in activation-induced apoptotic death in T cell hybridomas. *Science* 257:212 (1992).

15

TAKING OUT THE IMMUNE RESPONSE

The Roles of Fas–Ligand (CD95L) in Immune Regulation

Douglas R. Green,[1] Brian Tietz,[1] Thomas A. Ferguson,[2] and Thomas Brunner[1]

[1]Division of Cell. Immunol.
La Jolla Institute for Allergy and Immunology
San Diego, California
[2]Department of Ophthalmology
Washington University School of Medicine
St. Louis, Missouri

INTRODUCTION

Since the discovery of Fas (Apo-1/CD95) and its ligand (FasL), the rapid and profound apoptosis triggered by this receptor has held the imagination of researchers in the areas of cell biology and immunology. Antibodies to Fas are potent inducers of apoptosis in many different cell types (Yonehara et al., 1989, Trauth et al., 1989, Owen-Schaub et al., 1992), and the mechanism of this effect is the subject of extensive research (and some controversy). This overview, however, will deal with a different aspect of Fas-mediated apoptosis: its role in the regulation of immune responses.

One of the first indications that Fas/FasL interactions were potentially important in the regulation of immune responses was the identification of the *lpr* mutation as a defect in expression of Fas (Watanabe-Fukunaga et al., 1992). The defect, inhibition of Fas gene transcription due to a retroviral transposon insertion (Adachi et al., 1993), results in massive accumulation of B220$^+$, TCR$\alpha\beta^+$, CD4$^-$, CD8$^-$ T cells in lymph nodes and spleen and an age-accelerated production of autoantibody in some backgrounds. Similarly, point mutations in the gene for FasL, resulting in a FasL protein that does not trigger Fas, cause the *gld* mutant phenotype (Takahashi et al., 1994), which is basically the same as that of *lpr*.

By themselves, these observations demonstrate that Fas/FasL interactions are critically important in the maintenance of normal immune homeostasis. We will return to a consideration of the disease states caused by mutations in these genes as we try to make sense of the function of FasL and its receptor in the regulation of the immune system.

Programmed Cell Death, edited by Shi *et al.*
Plenum Press, New York, 1997

ACTIVATION-INDUCED APOPTOSIS IN T CELLS

The clonal selection theory of the immune system predicted that specific stimulation of lymphocytes via their antigen receptors should lead to either deletion or proliferation of the cell, depending upon its stage of development. This fundamental concept remains one of the strongest models we have for the development and function of the immune system. More recently, the concept of clonal deletion has been expanded to those of negative selection (refering to clonal deletion of immature lymphocytes) and peripheral deletion (refering to clonal deletion of mature lymphocytes). But the concept stays the same: under some circumstances, antigen-specific activation of a lymphocyte can result in its elimination from the immune system.

Conceptually it is a small step from the idea that specific activation eliminates a cell to the notion of activation-induced cell death (AICD). Activation of a lymphocyte can stimulate it to die. While hardly remarkable, it was nevertheless important that this death was recognized as being apoptotic. Thus, activation of thymocytes (Smith et al., 1989), T cell hybridomas (Shi et al., 1990, Ucker et al., 1989), or previously stimulated mature T cells (Russell, et al., 1991) was found to induce apoptosis. More recently, activation-induced apoptosis in B cells and B cell lines was demonstrated as well (Valentine & Licciardi, 1992).

The first indication that some forms of activation-induced apoptosis might proceed via Fas/FasL interactions came from the studies of Russell and colleagues, who demonstrated that apoptosis induced by restimulation of previously activated T cells does not occur in cells from *lpr* (Russell et al., 1993) or *gld* (Russell & Wang, 1993) mice. Other studies established that T cells can express both Fas and FasL, and it was clear that activation could induce not only functional FasL expression (Rouvier et al., 1993) but also a delayed susceptibility to Fas-mediated apoptosis (Owen-Schaub et al., 1992, Klas et al., 1993). Subsequently, Lynch and colleagues (Alderson et al., 1995) showed that competitive inhibition of Fas/FasL interactions inhibited activation-induced apoptosis in restimulated human T cells. These studies clearly indicated that activation-induced apoptosis *can* proceed via Fas/FasL interactions.

Do they always? Studies in *lpr* mice (Singer & Abbas, 1994) and in Fas null animals (Adachi et al., 1996) showed that Fas is not required for negative selection of thymocytes, although one recent study suggests that Fas/FasL interactions may participate in activation-induced apoptosis of thymocytes under some conditions (Fisher et al., 1996). Similarly, activation-induced apoptosis in B cell lines proceeds independently of Fas and its ligand (see D. Scott, et al in this volume). Further, activation-induced apoptosis in restimulated T cells (Zheng et al., 1995) and peripheral deletion in vivo (Sytwu et al., 1996) can proceed in the absence of Fas/FasL interactions, via a TNF/TNFR-dependent cell death.

One model system for the study of activation-induced apoptosis involves the activation of T cell hybridomas. Ashwell and colleagues first observed that ligation of the T cell receptor on T cell hybridomas can trigger cell cycle arrest and cell death (Ashwell et al., 1987). Shi, et al (Shi et al., 1990) showed that this occurs via apoptosis, which was analyzed in more detail by these and other authors (Mercep et al., 1989, Ucker et al., 1989, Shi et al., 1989, Odaka et al., 1990). This phenomenon was extensively analyzed in terms of signals and molecular events necessary for the effects, and theories relating activation, cell cycle progression, and apoptosis were offered to account for this process (e.g., Green, et al., 1994). However, it was subsequently demonstrated that activation-induced apoptosis in T cell hybridomas (Brunner et al., 1995, Ju et al., 1995) and Jurkat cells (Dhein et al., 1995) is dependent upon induction of FasL and interaction with Fas on the surface of these cells, a process that can occur even on a single cell.

Activation-induced apoptosis in T cell hybridomas remains a useful model for studying the regulation of FasL expression following T cell activation, and much of our thinking about Fas/FasL interactions is based on the use of such models. However as we will see, these useful models may be insufficient to fully explain the role of FasL expression in immune homeostasis.

ACTIVATION SIGNALS LEADING TO FasL-MEDIATED APOPTOSIS IN T CELLS

Activation-induced cell death can potentially be controlled at the level of Fas ligand expression, Fas expression, and Fas signal transduction. The majority of conditions which block activation-induced cell death do so by blocking expression of Fas ligand, with partial or no effect on Fas expression (see below). It therefore seems that expression of Fas ligand is more stringently regulated than expression of Fas. As such, the regulation of Fas ligand expression is currently the subject of intense investigation. However, in a later section we will also consider one case in which the regulation of activation-induced apoptosis is not via control of Fas–ligand expression but rather resistance to Fas-induced apoptosis.

The process of activation-induced apoptosis is summarized in Figure 1. This phenomenon proceeds via ligation of the T cell receptor and signal transduction events culminating in expression of the Fas–ligand gene. In one current model for T-cell activation, during stimulation of a T cell under physiological conditions, the coreceptor extracellular domain engages MHC, thereby anchoring it into the TCR complex. The short intracellular domain of the coreceptor provides a binding site for the Src family kinase Lck, which then phosphorylates antigen recognition activation motifs on the intracellular domains of the CD3 and TCRξ subunits (Iwashima et al., 1994). Phosphorylation of TCRξ is then thought to recruit the Syk family kinase ZAP-70 via its SH2 domain (Chan et al., 1992). This initiates the activation of numerous kinases, including Fyn, Yes, and Syk, not all of which are present in all T-cells. Thus, at the engaged TCR complex, there forms an extensive network of kinases and phosphorylation. As would be expected, Herbimycin A, an inhibitor of Src family protein tyrosine kinases (PTKs), blocks AICD and Fas ligand upregulation in response to TCR ligation (Anel et al., 1994), and this drug also blocks TCR-mediated AICD in T cell hybridomas (unpublished observations). It does not, however, block upregulation of Fas ligand in response to stimulation with PMA/ionomycin, indicating that no Herbimycin A-sensitive Src family PTKs are involved downstream of PKC in Fas ligand regulation. A non-Src family PTK must however be involved downstream of PKC, as Genistein, a broad inhibitor of PTKs, blocks expression of Fas ligand in response to either TCR ligation or PMA/ionomycin.

The immunosuppressive drugs CsA and FK506 exert their immunosuppressive effects, at least in part, by preventing Fas ligand upregulation, with no effect detectable on Fas mRNA level, or on Fas signal transduction in response to anti-Fas antibodies (Brunner et al., 1996). CsA and FK506 affect Fas ligand expression by binding to the cellular proteins cyclophilin and FKBP, respectively, facilitating their binding to and inactivation of calcineurin (Liu et al., 1991). Calcineurin is known to activate NF-κB and NFAT, making them potential activators of Fas ligand transcription. Indeed, a putative κB site has been identified in the Fas ligand promoter. Although no NFAT response element has yet been found, that does not rule out NFAT involvement, as only 450 bp of the Fas ligand promoter immediately upstream of the start codon have been reported (Takahashi et al., 1994).

Another transcription factor required for AICD in some cases is the orphan steroid receptor Nur77. Its expression and activity following TCR activation has been shown to

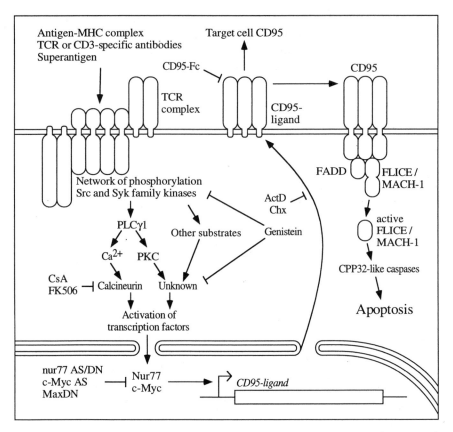

Figure 1. Regulation of activation-induced FasL (CD95L) expression. TCR activation by antigen-presenting cells presenting antigen, antibodies, or superantigen leads to early activation of PTKs of the Src and Syc family. This in turn leads to elevation of intracellular Ca^{2+} and the activation of the CsA/FK506-sensitive phosphatase calcineurin. In parallel, TCR activation induces activation of PKC and other genistein/herbimycin-sensitive kinases. Both pathways appear to lead to the activation of transcription factors resulting in FasL transcription. De novo synthesized FasL can bind to either Fas (CD95) on a neighboring target cell or on the same cell, engaging the Fas signaling pathway leading to apoptotic cell death. TCR, T cell receptor; CsA, cyclosporin A; PKC, protein kinase C; AS, antisense oligonucleotides; DN, dominant negative; Act D, actinomycin D; Chx, cycloheximide; CD95-Fc, CD95/Fas–human IgG fusion protein.

be required for AICD in T cell hybridomas using both antisense (Liu et al., 1994) and dominant negative (Woronicz et al., 1994) approaches. Nur77 expression appears to be insufficient to confer increased Nur77 DNA binding, with a requirement for further modification to activate it. Activation of Nur77 is likely to be mediated by a protein downstream of calcineurin, as CsA blocks Nur77 DNA binding activity without inhibiting its expression (Yazdanbakhsh et al., 1995). Interestingly, despite its requirement for AICD in T cell hybridomas, AICD of neither thymocytes or peripheral T cells was impaired in nur77 knockout mice (Lee et al., 1995). This may indicate that Nur77 has a functional homolog *in vivo* which is lacking in T cell hybridomas. Alternatively, it may simply reflect different requirements for death in T cell hybridomas and T cells or thymocytes.

A number of steroid hormones affect the expression of Fas ligand. Glucocorticoids can induce apoptosis in T cells when used alone (Iseki et al., 1991, Zacharchuk et al., 1991), but can also block AICD in both T cell hybridomas and thymocytes (Iseki et al.,

1991, Iwata et al., 1991, Zacharchuk et al., 1990). Thus, the mechanisms for glucocorticoid-induced apoptosis and AICD are mutually exclusive. In T cells, glucocorticoids block AICD by preventing upregulation of Fas ligand (Yang et al., 1995). Similarly, activation of the Retinoid X Receptor (RXR) and Retinoic Acid Receptor (RAR) by agonists like 9-cis retinoic acid can block activation-induced cell death (Bisonnette et al., 1995). Inhibition of AICD by RXR and RAR requires activation of both receptors, as pan-agonists like 9-cis retinoic acid can inhibit AICD, while RAR- or RXR-selective agonists are ineffective at blocking AICD unless used together, in which case they can block AICD. Like glucocorticoids, 9-cis retinoic acid inhibits activation-induced cell death by blocking Fas ligand upregulation.

ACTIVATION-INDUCED APOPTOSIS AND CELL CYCLE REGULATORS

The control of signals leading to AICD, while important for our understanding of this phenomenon, could be viewed as similar to the study of signaling leading to cytokine gene expression, especially since FasL is related to known cytokines such as TNF. There is one aspect of AICD, however, that suggests that the regulation of FasL expression and apoptosis in the cells may be more complicated (and, perhaps, more interesting). This is the apparent role of the cell cycle in the regulation of AICD.

Ashwell and colleagues (Ashwell et al., 1987) were the first to observe that activation of T cell hybridomas leads to a cell cycle arrest, although the possible connection to AICD was not made. Lenardo and colleagues (Boehme & Lenardo, 1993) showed that inhibitors of cell cycle progression inhibit AICD in activated T cells, and subsequently others (Radvanyi et al., 1996a) have suggested that AICD in these cells occurs in a cell cycle-restricted fashion, predominantly in late G1 or early S.

In contrast, Fotedar et al (Fotedar et al., 1995) have shown that activation-induced apoptosis in T cell hybridomas occurs in late G2, prior to expression of M-phase markers. They observed this in synchronized cells, where cell death did not occur until the cells completed S phase. In addition, they found that antisense oligonucleotides to cyclin B blocked AICD, suggesting a direct connection of this phenomenon to regulation of the cell cycle.

Since AICD in these systems proceeds via FasL expression and interaction with Fas, this cell cycle regulation of AICD can only be explained in two ways: either expression of FasL or susceptibility to Fas-induced apoptosis must be under cell cycle control. We have so far been unable to show any differences in susceptibility to Fas-mediated apoptosis in the cell cycle (unpublished results), although one study suggests that Fas-mediated apoptosis is dependent upon the function of cdc2 kinase (Yao et al., 1996). In contrast, our preliminary results suggest that FasL may indeed be under cell cycle control, based on two observations (Brunner, et al., unpublished results). First, the cyclin B antisense oligonucleotides described by Fotedar, et al (Fotedar et al., 1996) effectively block expression of FasL mRNA and FasL function following activation, thus providing a mechanism for this effect. Secondly, cell surface staining of FasL following activation of T cell hybridomas predominantly labels cells in G2/M of the cell cycle. Together with the results of Fotedar, et al. (Fotedar et al., 1996), these findings suggest the remarkable idea that the expression and function of FasL is under cell cycle control.

Another intiguing aspect of FasL control is its possible regulation by the cell cycle regulator c-Myc. Expression of c-myc is constitutive in T cell hybridomas, and this is required for AICD, as antisense oligonucleotides to c-myc (Shi et al., 1992) can block activation-induced apoptosis in these cells. Further, expression of a dominant negative form of Max (MaxRX), which contains the Myc leucine zipper helix-loop-helix (LZ-HLH), and thus forms nonfunctional dimers with Max, functions to block AICD in T cell hybridomas (Bissonette, et al., 1994). This inhibition is reversed by coexpression of another mutant, MycRX, which contains the Max LZ-HLH, resulting in a functional MycRX/MaxRX heterodimer. These studies show that functional Myc/Max heterodimer formation is necessary for AICD. c-Myc is required for AICD per se, not simply T cell activation, as IL-2 production following activation was not affected by c-myc antisense oligonucleotides or dominant negative constructs.

c-Myc induces or promotes apoptosis in other systems as well (Wurm et al., 1986, Wyllie et al., 1987, Askew et al., 1991, Evan et al., 1992). In these systems, Myc-mediated expression of p53 (Hermeking & Eick, 1994) or ornithine decarboxylase (Askew et al., 1991) appear to be important in mediating the effects of Myc on cell death. In AICD, c-Myc function appears to be required for upregulation of Fas ligand, as c-myc antisense oligonucleotides block Fas ligand expression following activation (T. Brunner, N. J. Yoo, D. R. Green, unpublished results). Studies are currently in progress to determine whether a putative Myc/Max binding site in the FasL promoter (T. Brunner, S. Khasibhatla, D.R. Green, unpublished results) is important for FasL expression. If so, then FasL may be one of a small number of genes whose expression is controlled by c-Myc.

Thus, the regulation of FasL expression appears to be more complicated than one might have expected based only on T cell activation signals and similarity to known cytokines. Requirements for Myc and perhaps cyclin-dependent kinases for optimal FasL expression may manifest as a requirement for T cells (and perhaps other cells? see below) to be in cycle before functional FasL appears, leading to suicide of the cell or killing of other Fas$^+$ cells.

CONTROL OF SUSCEPTIBILITY TO Fas-MEDIATED APOPTOSIS IN THE REGULATION OF IMMUNE RESPONSES

Much of the research on the regulation of AICD has focused on the expression of FasL, as we've discussed. It is also clear, however, that susceptibility to Fas-mediated apoptosis is also controlled, and this is also important in immune homeostasis.

Activated T lymphocytes have increased levels of Fas expression, but susceptibility to Fas-mediated apoptosis usually doesn't appear in these cells until several days after the appearance of Fas (Owen-Schaub et al., 1992, Klas et al., 1993). Similarly some cell types express Fas but do not undergo apoptosis upon ligation of this receptor. In both cases, treatment of cells with inhibitors of macromolecular synthesis can reveal susceptibility to Fas-mediated apoptosis, and thus the resistance is an active one (as opposed to the absence of functional signaling molecules). The mechanism of this resistance is not known, but there are at least two candidates. Bcl-xL can inhibit Fas-induced apoptosis in T lymphocytes, and some treatments that produce resistance to AICD elevate expression of this molecule (Kinoshita et al., 1995, Radvanyi et al., 1996b). Another candidate is a tyrosine phosphatase, FAP-1, which can produce partial resistance to Fas-mediated apoptosis when expressed in Jurkat cells (Sato et al., 1995). How either of these molecules acts to block Fas-mediated apoptosis is unknown.

The type of immune response that may be generated by a particular antigenic insult is determined in part by the profile of cytokines that are produced by activated T cells. Different sets of cytokines correspond to different types of immune responses, and during the course of an immune response, the profile of cytokines produced tends to skew towards one or another set. This polarized response represents the maturation of different CD4$^+$ T cell subsets, predominantly those of Th1 and Th2 cells. On first pass, Th1 cells tend to promote cell mediated immune responses, while Th2 cells promote humoral and atopic immunity.

Lines of Th1 versus Th2 cells have been examined for susceptibility to AICD, and one group has concluded that expression of FasL occurs predominantly in Th1 cells but not in Th2 cells, thus resulting in a relative resistance of the latter to AICD (Ramsdell et al., 1994). Others, however, have obtained different results, finding that Th2 lines can express FasL (Suda et al., 1995). Recently, together with Swain and colleages (Zhang et al., 1997), we have examined the susceptibility of short-term cultured, polarized T cells (showing properties of Th1 or Th2 cells) to AICD. We found that although both Th1 and Th2 cells express functional FasL following activation, and both express Fas, only Th1 cells undergo AICD. Even when both cell types are mixed together, all of the activation-induced apoptosis occurs in the Th1 population. Although no differences in expression of Bcl-2 or Bcl-x were observed, we did find a dramatic difference in expression of FAP-1. The Th2 cells expressed FAP-1 while the Th1 cells did not. It is tempting to speculate that this differential expression of FAP-1 is responsible for the resistance of Th2 cells to AICD.

Whatever the mechanism of this resistance, it is easy to see that Th2 cells should have an advantage over Th1 cells when both are activated in vivo, since the Th1 cells are susceptible to AICD. Under other circumstances in which T cells become exposed to FasL (see below), this difference in susceptibility to Fas-mediated apoptosis should manifest as a skewing towards Th2 dominance. In turn, the regulatory influences of Th2 cells will then contribute to the spectrum of immunoregulation during the response. Thus, these observations will help us to understand how two important regulatory functions, FasL-induced apoptosis and Th subset cross-talk, are coordinated to control immune responses.

Fas LIGAND CONTROL OF IMMUNE PRIVILEGE

Long before we had any understanding of the nature of the immune response, it was recognized that some tissues are "privileged" in that they will support the growth of cells or parasites that will not grow elsewhere in the body. Later it became clear that these privileged sites were places where immune responses were not induced. The first such site recognized was the anterior chamber of the eye. Subsequently, other tissues such as the testes were recognized as being immunologically privileged. While the phenomenon of immunologic privilege has been known for over 100 years, it is only recently that we have gained insights into how this privilege is brought about.

The protection of immunologically privileged tissues from destructive inflammatory responses is dependent upon the presence of functional FasL expressed on those tissues. For example, we demonstrated that the surfaces surrounding the anterior chamber of the eye express functional FasL, capable of inducing apoptosis in Fas$^+$ cells (Griffith, et al., 1995). Injection of HSV-1 into the anterior chamber of the eye of normal mice resulted in infiltration of inflammatory cells and their subsequent apoptosis, leaving the eye intact. Injection of HSV-1 into *gld* mice, however, resulted in an extensive, destructive inflammatory response. Bone marrow chimera experiments confirmed that the peripheral tissues, not bone marrow derived tissues, are the source of FasL responsible for inducing apoptosis in infiltrating inflammatory cells.

Another manifestation of immunological privilege is resistance to allograft rejection. Testis grafts can survive when transplanted under the kidney capsule of allogeneic animals (Bellgrau, et al., 1995). This survival is dependent on expression of functional Fas ligand, as grafts from *gld* donor mice were rejected. More recently, it was similarly demonstrated that human and murine corneas express functional FasL, and acceptance of murine corneal allografts was shown to be dependent on the expression of this molecule (Stuart et al., 1997).

The remarkable observation that FasL is necessary for resistance to graft rejection raises the possibility of engineering tissues to allow acceptence of allografts. Lau, et al (Lau et al., 1996) embedded pancreatic islets in myoblasts that had been transfected to express FasL, and these were implanted into mice allogeneic for the islets. The presence of FasL in the syngeneic myoblasts dramatically delayed rejection of the graft.

The cells that enter an immunologically privileged site and undergo apoptosis do so via Fas-mediated cell death. This was formally demonstrated by engrafting normal mice with *lpr* bone marrow, then challenging the eye with virus as mentioned above (Griffith, et al., 1996). The infiltrating cells, lacking Fas, failed to undergo apoptosis. Thus, immune privilege proceeds via FasL/Fas interactions resulting in the apoptotic death of infiltrating inflammatory cells. This is summarized in Figure 2, in which activation-induced apoptosis of T cells is also shown for comparison.

Since Th2 cells can be resistant to the effects of FasL (see above), we might expect that antigenic exposure in a site such as the eye promotes Th2 generation while destroying Th1 cells. Some support for this idea exists. Antigen challenge in the eye leads to the condition of anterior chamber associated immune deviation (ACAID), where cell mediated immune responses are suppressed but humoral responses proceed. Manipulations affecting the generation of Th2 cells can prevent the generation of ACAID (Geiger & Sarvetnick, 1994). The phenomenon is somewhat more complicated, however, since we have recently found that ACAID depends in part on the presence of apoptotic cells together with antigen, not merely the deletion of Th1 cells (Griffith et al., 1996). The complex relationships between Fas/FasL interactions, apoptosis, and immune deviation are only beginning to be unraveled.

Fas ligand is expressed in several other immune privileged sites. In developing embryos, Fas ligand mRNA was detected by *in situ* hybridization in salivary gland, brain, and spinal cord, as well as thymus (French et al., 1996). In adult mice, Fas ligand mRNA was detected by RNAse protection assay in lung, small and large intestines, liver, adrenal gland, seminal vesicle, testis, ovary and uterus, as well as thymus and spleen (French et al., 1996). Of these, the brain, adrenal gland, testis, ovary, and uterus are known immune privileged sites. Fas ligand may therefore play a protective role in these tissues, a question which is currently the subject of investigation. Fas ligand does indeed appear to have a mildly protective effect in allogeneic adrenal gland grafts (B. Tietz, T. Brunner and D. R. Green, unpublished data).

Could Fas ligand expression by tissues be a global mechanism for determining the extent of the immune response allowed in any given tissue? It has been clearly demonstrated that some privileged tissues protect themselves from destructive inflammatory responses by constitutively expressing Fas ligand (see Figure 2). In other tissues, however, the threshold of damage which can be tolerated may vary considerably. It therefore seems possible that any tissue might express Fas ligand during the course of an immune response if the response becomes so damaging as to be self-defeating. This idea is shown schematically in Figure 2. For example, the liver is less sensitive to inflammatory responses than the eye or the testes, but an extensive, sustained immune response in the liver may cause so much damage as to result in hepatic failure and cirrhosis. Is it then possible that tissues

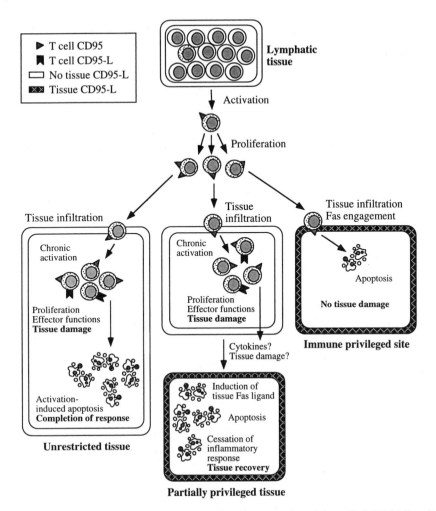

Figure 2. Thresholds of damage or inflammatory cytokines for expression of tissue FasL (CD95 ligand). Left: lymphocytes infiltrating into tissues which allow an unrestricted inflammatory response proliferate in the tissue and fulfill effector fuctions, causing tissue damage. The immune response is self-regulating and is terminated by activation-induced apoptosis. Center: lymphocytes infiltrating into a partially privileged tissue with a lower threshold for damage proliferate in the tissue and fulfill effector fuctions, causing tissue damage. The tissue participates in the regulation of the immune response by expressing CD95 ligand in a cytokine- or damage-dependent manner, thereby causing the lymphocytes to undergo apoptosis and protect the tissue. The ability of such tissues to inducibly express CD95/Fas–ligand is hypothetical. Right: lymphocytes infiltrating into an immune privileged site encounter CD95 ligand immediately and undergo apoptosis, preventing damage to the tissue. CD95-L, CD95/Fas ligand.

not recognized as immune privileged sites may begin to express Fas ligand if an immune response exceeds a tolerable level? Different tissues may have different thresholds at which they begin to express Fas ligand as a global mechanism for the limitation of immune responses. Fas ligand expression in such a mechanism could be responsive to the concentration of inflammatory cytokines. Immune privileged sites such as the eye and testes may then represent extreme cases with zero tolerance for inflammation, while durable tissues such as skin which provide the first line of defense against pathogens may allow unrestricted immune responses. This speculative model for an interaction between the extent of an immune response and the level of FasL expression in different tissues would have far reaching consequences for understanding immune regulation. Several approaches to testing this model are currently underway.

ACKNOWLEDGMENTS

Our research on activation-induced apoptosis and the regulation of Fas-ligand expression is supported by grants from the NIH to DRG and TAF. The authors thank Dr. Tom Griffith for discussions and for providing "FasL" beer.

REFERENCES

Adachi, M., Suematsu, S., Suda, T., Watanabe, D., Fukuyama, H., Ogasawara, J., Tanaka, T., Yoshida, N., Nagata, S. (1996). Enhanced and accelerated lymphoproliferation in Fas-null mice. Proc. Natl. Acad. Sci. USA93:2131–2136.

Adachi, M., Watanabe-Fukunaga, R., Nagata, S. (1993). Aberrant transcription caused by the insertion of an early transposable element in an intron of the Fas antigen gene of *lpr* mice. Proc. Natl. Acad. Sci. USA90(5):1756–1760.

Alderson, M. R., Tough, T. W. Davis-Smith, T., Braddy, S., Falk, B., Schooley, K. A., Goodwin, R. G., Smith, C. A., Ramsdell, F., Lynch, D. H. (1995). Fas ligand mediates activation-induced cell death in human T lymphocytes. J. Exp. Med. 181:71–77.

Anel, A., Buferne, M., Boyer, C., Schmitt-Verhulst, A. M., Golstein, P. (1994). T cell receptor-induced Fas ligand expression in cytotoxic T lymphocyte clones is blocked by protein tyrosine kinase inhibitors and cyclosporin A. Eur. J. Immunol. 24:2469–2476.

Ashwell, J. D., Cunningham, R. E., Noguchi, P. D., Hernandez, D. (1987). Cell growth cycle block of T cell hybridomas upon activation with antigen. J. Exp. Med. 165:173–194.

Askew, D. S., Ashmun, R. A., Simmons, B. C., Cleveland, J. L. (1991). Constitutive c-myc expression in an IL-3-dependent myeloid cell line suppresses cell cycle arrest and accelerates apoptosis. Oncogene 6:1915–1922.

Bellgrau, D., Gold, D., Selawry, H., Moore, J., Franzusoff, A., Duke, R. C. (1995). A role for CD95 ligand in preventing graft rejection. Nature 377:630–632.

Bisonnette, R. P., Brunner, T., Lazarchik, S. B., Yoo, N. J., Boehm, M. F., Green, D. R., Heyman, R. A. (1995). 9-cis Retinoic Acid Inhibition of Activation-Induced Apoptosis Is Mediated via Regulation of Fas Ligand and Requires Retinoic Acid Receptor and Retinoid X Receptor Activation. Mol. Cell. Biol.15:5576–5585.

Bisonnette, R. P., McGahon, A., Mahboubi, A., Green, D. R. (1994). Functional Myc-Max Heterodimer Is Required for Activation-induced Apoptosis in T Cell Hybridomas. J. Exp. Med. 180:2413–2418.

Boehme, S. A., Lenardo, M. J. (1993). Propriocidal apoptosis of mature T lymphocytes occurs at S phase of the cell cycle. Eur. J. Immunol. 23:1552–1560.

Brunner, T., Mogil, R. J., LaFace, D., Yoo, N. J., Mahboubi, A., Echeverri, F., Martin, S. J., Force, W. R., Lynch, D. H., Ware, C. F., Green, D. R. (1995). Cell-autonomous Fas (CD95)/Fas-ligand interaction mediates activation-induced apoptosis in T-cell hybridomas. Nature 373:441–444.

Brunner, T., Yoo, N. J., LaFace, D., Ware, C. F., Green, D. R. (1996). Activation-induced cell death in murine T cell hybridomas. Differential regulation of Fas (CD95) versus Fas ligand expression by cyclosporin A and FK506. Int. Immunol. 8:1017–1026.

Chan, A. C., Iwashima, M., Turck, C. W., Weiss, A. (1992). ZAP-70: a 70 kd protein-tyrosine kinase that associates with the TCR zeta chain. Cell 71, 649–662.

Dhein, J., Walczak, H., Baumler, C., Debatin, K. M., Krammer, P. H. (1995). Autocrine T-cell suicide mediated by APO-1/(Fas/CD95). Nature 373:438–441.

Evan, G. I., Wyllie, A. H., Gilbert, C. S., Littlewood, T. D., Land, H., Brooks, M., Waters, C. M., Penn, L. Z., Hancock, D. C. (1992). Induction of apoptosis in fibroblasts by c-myc protein. Cell 69:119–128.

Fotedar, R., Flatt, J., Gupta, S., Margolis, R. L., Fitzgerald, P., Messier, H., Fotedar, A. (1995). Activation-induced T-cell death is cell cycle dependent and regulated by cyclin B. Mol. Cell. Biol.15:932–942.

French, L. E., Hahne, M., Viard, I., Radlgruber, G., Zanone, R., Becker, K., Müller, C., Tschopp, J. (1996). Fas and Fas Ligand in Embryos and Adult Mice: Ligand Expression in Several Immune-privileged Tissues and Co-expression in Adult Tissues Characterized by Apoptotic Cell Turnover. J. Cell Biol. 133:335–343.

Geiger, K., Sarvetnick, N. (1994) Local production of IFN-gamma abrogates the intraocular immune privilege in transgenic mice and prevents the induction of ACAID. J Immunol 153: 5239–5246.

Green, D.R., A. Mahboubi, W. Nishioka, S. Oja, F. Echeverri, Y. Shi, J. Glynn, J. Ashwell, and R. Bissonnette (1994) Promotion and inhibition of activation-induced apoptosis in T cell hybridomas by oncogenes and related signals. Immunol. Rev. 142: 321–342.

Griffith, T. S., Brunner, T., Fletcher, S. M., Green, D. R., Ferguson, T. A. (1995). Fas Ligand-Induced Apoptosis as a Mechanism of Immune Privilege. Science 270:1189–1192.

Griffith, T. S., Yu, X., Herndon, J. M., Green, D. R., Ferguson, T. A. (1996). CD95-Induced Apoptosis of Lymphocytes in an Immune Privileged Site Induces Immunological Tolerance. Immunity 5:7–16.

Hermeking, H., Eick, D. (1994) Mediation of c-Myc-induced apoptosis by p53. Science 265: 2091–2093.

Iseki, R., Kudo, Y., Iwata, M. (1991). Early mobilization of Ca2+ is not required for glucocorticoid-induced apoptosis in thymocytes. J. Immunol. 151:5198–5207.

Iseki, R., Mukai, M., Iwata, M. (1991). Regulation of T lymphocyte apoptosis. Signals for the antagonism between activation- and glucocorticoid-induced death. J. Immunol. 147:4286–4292.

Iwashima, M., Irving, B. A., van Oers, N. S., Chan, A. C., Weiss, A. (1994). Sequential interactions of the TCR with two distinct cytoplasmic tyrosine kinases. Science 263, 1136–1139.

Iwata, M., Hanaoka, S., Sato, K. (1991). Rescue of thymocytes and T cell hybridomas from glucocorticid-induced apoptosis by stimulation via the T cell receptor/CD3 complex: a possible in vitro model for positive selection of the T cell repertoire. Eur. J. Immunol. 21:643–648.

Ju, S. T., Panka, D. J., Cui, H., Ettinger, R., el-Khatib, M., Sherr, D. H., Stanger, B. Z., Marshak-Rothstein, A. (1995). Fas(CD95)/FasL interactions required for programmed cell death after T-cell activation. Nature 373:444–448.

Kinoshita, T., Yokota, T., Arai, K., Miyajima, A. (1995). Regulation of Bcl-2 expression by oncogenic Ras protein in hematopoietic cells. Oncogene 10:2207–2212.

Klas, C., Debatin, K. M., Jonker, R. R., Krammer, P. H. (1993). Activation interferes with the APO-1 pathway in mature human T cells. Int. Immunol. 5:625.

Lau, H. T., Yu, M., Fontana, A., Stoeckert, C. J. Jr. (1996). Prevention of islet allograft rejection with engineered myoblasts expressing FasL in mice. Science 273:109–112.

Lee, S. L., Wesselschmidt, R. L., Linette, G. P., Kanagawa, O., Russell, J. H., Milbrandt, J. (1995). Unimpaired thymic and peripheral T cell death in mice lacking the nuclear receptor NGFI-B (Nur77). Science 269:532–535.

Liu, J., Farmer, J. D., Lane, W. S., Friedman, J., Weissman, I., Schreiber, S. L. (1991). Calcineurin is a common target of cyclophilin-cyclosporin A and FKBP-FK506 complexes. Cell 66:807–815.

Liu, Z. G., Smith, S. W., McLaughlin, K. A., Schwartz, L. M., Osborne, B. A. (1994). Apoptotic signals delivered through the T-cell receptor of a T-cell hybrid require the immediate-early gene nur77. Nature 367:281–284.

Mercep, M., Noguchi, P. D., Ashwell, J. D. (1989). The cell cycle block and lysis of an activated T cell hybridoma are distinct processes with different Ca2+ requirements and sensitivity to cyclosporin A. J. Immunol. 142:4085–4092.

Odaka, C., Kizaki, H., Tadakuma, T. (1990). T cell receptor-mediated DNA fragmentation and cell death in T cell hybridomas. J. Immunol. 144:2096–2101.

Owen-Schaub, L. B., Yonehara, S., Crump, W. L., Grimm, E. A. (1992). DNA fragmentation and cell death is selectively triggered in activated human lymphocytes by Fas antigen engagement. Cell. Immunol. 140:197–205.

Radvanyi, L. G., Shi, Y., Mills, G. B., Miller, R. G. (1996a). Cell cycle progression out of G1 sensitizes primary-cultured nontransformed T cells to TCR-mediated apoptosis. Cell. Immunol. 170:260–273.

Radvanyi, L. G., Shi, Y., Vaziri, H., Sharma, A., Dhala, R., Mills, G. B., Miller, R. G. (1996b) CD28 costimulation inhibits TCR-induced apoptosis during a primary T cell response. J. Immunol. 156:1788–1798.

Ramsdell, F., Seaman, M. S., Miller, R. E., Picha, K. S., Kennedy, M. K., Lynch, D. H. (1994). Differential ability of Th1 and Th2 T cells to express Fas ligand and to undergo activation-induced cell death. Int. Immunol. 6:1545–1553.

Rouvier, E., Luciani, M. F., Golstein, P. (1993). Fas involvement in Ca(2+)-independent T cell-mediated cytotoxicity. J. Exp. Med. 177:195–200.

Russell, J. H., Rush, B., Weaver, C., Wang, R. (1993) Mature T cells of autoimmune *lpr/lpr* mice have a defect in antigen-stimulated suicide. Proc. Natl. Acad. Sci. USA90:4409–4413.

Russell, J. H., Wang, R. (1993) Autoimmune *gld* mutation uncouples suicide and cytokine/proliferation pathways in activated, mature T cells. Eur. J. Immunol. 23:2379–2382.

Russell, J.H., C.L. White, D.Y. Loh, and P. Meleedy-Rey (1991) Receptor-stimulated death pathway is opened by antigen in mature T cells. Proc. Natl. Acad. Sci. USA 88: 2151–2155.

Sato, T., Irie, S., Kitada, S., Reed, J. C. (1995). FAP-1: a protein tyrosine phosphatase that associates with Fas. Science 268:411–415.

Shi, Y., Glynn, J. M., Guilbert, L. J., Cotter, T. G., Bissonnette, R. P., Green, D. R. (1992). Role for c-myc in Activation-Induced Apoptotic Cell Death in T Cell Hybridomas. Nature 257:212–214.

Shi, Y., Sahai, B. M., Green, D. R. (1989). Cyclosporin A inhibits activation-induced cell death in T cell hybridomas and thymocytes. Nature 339:625–626.

Shi, Y., Szalay, M. G., Paskar, L., Boyer, M., Singh, B., Green, D. R. (1990). Activation-induced cell death in T cell hybridomas is due to apoptosis. J. Immunol. 144:3326–3333.

Singer, G. G., Abbas, A. K. (1994). The fas antigen is involved in peripheral but not thymic deletion of T lymphocytes in T cell receptor transgenic mice. Immunity 1:365–371.

Smith, C. A., Williams, G. T., Kingston, R., Jenkinson, E. J., Owen, J. J. (1989). Antibodies to CD3/T-cell receptor complex induce death by apoptosis in immature T cells in thymic cultures. Nature 337:181–184.

Stuart, P. M., Griffith, T. S., Usui, N., Pepose, J., Yu, X., Ferguson, T. A. (1997). CD95 ligand (FasL)-induced apoptosis is necessary for corneal allograft survival. J. Clin. Invest., 99:396–402.

Suda, T., Okazaki, T., Naito, Y., Yokota, T., Arai, N., Ozaki, S., Nakao, K., Nagata, S. (1995). Expression of the Fas ligand in cells of T cell lineage. J. Immunol. 154:3806–3813.

Sytwu, H. K., Liblau, R. S., McDevitt, H. O. (1996). The roles of Fas/APO-1 (CD95) and TNF in antigen-induced programmed cell death in T cell receptor transgenic mice. Immunity 5:17–30.

Takahashi, T., Tanaka ,M., Brannan, C. I., Jenkins, N. A., Copeland, N. G., Suda, T., Nagata, S. (1994). Generalized lymphoproliferative disease in mice, caused by a point mutation in the Fas ligand. Cell 76:969–976.

Takahashi, T., Tanaka, M., Inazawa, J., Abe, T., Suda, T., Nagata, S. (1994). Human Fas ligand: gene structure, chromosomal location and species specificity. Int. Immunol. 6:1567–1574.

Trauth, B., Klas, C., Peters, A. M. J., Matzku, S., Moller, P., Falk, W., Debatin, K. M., Krammer, P. H. (1989). Monoclonal antibody-mediated tumor regression by induction of apoptosis. Science 245:301–305.

Ucker, D. S., Ashwell, J. D., Nickas, G. (1989). Activation-driven T cell death. I. Requirements for de novo transcription and translation and association with genome fragmentation. J. Immunol. 143:3461–3469.

Valentine, M. A., Licciardi, K. A. (1992). Rescue from anti-IgM-induced programmed cell death by the B cell surface proteins CD20 and CD40. Eur. J. Immunol. 22:3141–3148.

Watanabe-Fukunaga, R., Brannan, C. I., Copeland, N. G., Jenkins, N. A., Nagata, S. (1992). Lymphoproliferation disorder in mice explained by defects in Fas antigen that mediates apoptosis. Nature 356:314–317.

Woronicz, J. D., Calnan, B., Ngo, V., Winoto, A. (1994). Requirement for the orphan steroid receptor Nur77 in apoptosis of T-cell hybridomas. Nature 367:277–281.

Wurm, F. M., Gwinn, K. A., Kingston, R. E. (1986). Inducible overproduction of the mouse c-myc protein in mammalian cells. Proc. Natl. Acad. Sci. USA83:5414–5418.

Wyllie, A. H., Rose, K. A., Morris, R. G., Steel, C. M., Foster, E., Spandidos, D. A. (1987). Rodent fibroblast tumours expressing human myc and ras genes: growth, metastasis and endogenous oncogene expression. Br. J. Cancer56:251–259.

Yang, Y., Mercep, M., Ware, C. F., Ashwell, J. D. (1995). Fas and activation-induced Fas ligand mediate apoptosis of T cell hybridomas: inhibition of Fas ligand expression by retinoic acid and glucocorticoids. J. Exp. Med. 181: 1673–1682.

Yao, S. L., McKenna, K. A., Sharkis, S. J., Bedi, A. (1996). Requirement of p34cdc2 kinase for apoptosis mediated by the Fas/APO-1 receptor and interleukin 1beta-converting enzyme-related proteases. Cancer Res.56:4551–4555.

Yazdanbakhsh, K., Choi, J. W., Li, Y., Lau, L. F., Choi, Y. (1995). Cyclosporin A blocks apoptosis by inhibiting the DNA binding activity of the transcription factor Nur77. Proc. Natl. Acad. Sci. USA92:437–441.

Yonehara, S., Ishii, A., Yonehara, M. (1989). A cell-killing monoclonal antibody (anti-Fas) to a cell surface antigen co-downregulated with the receptor of tumor necrosis factor. J. Exp. Med. 169:1747–1756.

Zacharchuk, C. M., Mercep, M., Chakraborti, P. K., Simon, S. S., Ashwell, J. D. (1990). Programmed T lymphocyte death. Cell activation- and steroid-induced pathways are mutually antagonistic. J. Immunol. 145:4037–4045.

Zacharchuk, C. M., Mercep, M., June, C. H., Weissman, A. M., Ashwell, J. D. (1991). Variations in thymocyte susceptibility to clonal deletion during ontogeny. Implications for neonatal tolerance. J. Immunol. 147:460–465.

Zhang, X., Brunner, T., Carter, L., Dutton, R., W., Rogers, P.,, Sato, T., Reed, J., Green, D. R., Swain, S. L. (1997). Unequal death in Th1 and Th2 effectors: Th2 effectors express FAP-1 and are resistant to rapid activation induced cell death. J. Exp. Med. 185:1837–1849.

Zheng, L., Fisher, G., Miller, R. E., Peschon, J., Lynch, D. H., Lenardo, M. J. (1995). Induction of apoptosis in mature T cells by tumour necrosis factor. Nature 377:348–351.

16

CD95 (Fas) LIGAND

A Mediator of Cytotoxic T Lymphocyte-Mediated Apoptosis and Immune Privilege

Richard C. Duke,[1,2,3] Paul B. Nash,[4] Mary S. Schleicher,[5] Cynthia Richards,[6] Jodene Moore,[2] Evan Newell,[5,7] Alex Franzusoff,[8] and Donald Bellgrau[2,6]

[1]Division of Medical Oncology
Department of Medicine
University of Colorado Health Sciences Center
[2]Department of Immunology, UCHSC
[3]CERES Pharmaceuticals, Ltd.
Denver, Colorado
[4]Division of Infectious Disease
Department of Medicine, UCHSC
[5]Immunology Core
University of Colorado Cancer Center, UCHSC
[6]Barbara Davis Center for Childhood Diabetes, UCHSC
[7]McGill University
Montreal, Quebec, Canada
[8]Department of Cellular and Structural Biology, UCHSC
Denver, Colorado 80262

1. CD95 AND ITS LIGAND: DOWN-REGULATION OF IMMUNE RESPONSES

CD95 (Fas/APO-1) is a 45 kD cell surface glycoprotein and member of the tumor necrosis factor receptor superfamily (Nagata and Golstein 1995; Itoh et al. 1991; Oehm et al. 1992; Watanabe-Fukunaga et al. 1992a; Smith et al. 1994). Although CD95 is expressed in many tissues including liver, heart, gut, skin and ovaries (Watanabe-Fukunaga et al. 1992b; Leithauser et al. 1993), its major biological role appears to be in the regulation of immune responses (Nagata and Golstein 1995; Cohen and Eisenberg 1992; Vignaux and Golstein, 1994). *Lpr* mice which lack the ability to express functional CD95 accumulate large numbers of abnormal T and B cells in their peripheral lymphoid organs and develop autoimmune disease (Cohen and Eisenberg 1992; Cohen and Eisenberg 1991; Roths et al. 1984). The abnormal lymphocytes in these mice express markers found on activated lymphocytes and it appears that autoimmunity develops as a result of an inability to eliminate autoreactive T and B cells.

Antibodies to CD95 induce apoptosis in a variety of cell lines derived from lymphoid leukemias (Trauth et al. 1989; Yonehara et al. 1989; Debatin et al. 1990). In contrast, resting lymphocytes express low levels of CD95 and are not killed by anti-CD95 antibodies (Miyawaki et al. 1992; Owen-Schaub et al. 1992; Alderson et al. 1994; Daniel and Krammer 1994). Within 24 hours following antigen stimulation, however, T cells upregulate CD95 expression and after an additional two or three days will die if CD95 is crosslinked (Miyawaki et al. 1992; Owen-Schaub et al. 1992). These observations, together with the observed characteristics of *lpr* mice, suggest that expression of functional CD95 on activated lymphocytes targets them for destruction. CD95-mediated killing would thus limit immune responses in general and provide a safeguard against emergence of autoreactive lymphocytes (Nagata and Golstein 1995; Alderson et al. 1995; Strasser 1995).

In vivo, crosslinking of CD95 is mediated by CD95L, a 40 kD, type II membrane glycoprotein related to tumor necrosis factor (Nagata and Golstein 1995; Rouvier et al. 1993; Suda et al. 1993; Suda and Nagata 1994; Takahashi et al. 1994a; Takahashi et al. 1994b). *Gld* mice express a nonfunctional form of CD95L due to a single amino acid substitution in an extracellular region, and exhibit a disease identical to *lpr* mice. In contrast to *lpr* mice, however, the accumulation of autoreactive lymphocytes in *gld* mice can be reversed by adoptive transfer of lymphocytes from normal mice which are capable of expressing CD95L (Sobel et al. 1993; Sobel et al. 1995) suggesting that the autoimmune *gld* lymphocytes express functional CD95.

By Northern hybridization studies, CD95L mRNA has been found to be highly expressed in spleen and lymph node populations enriched for lymphocytes (Suda et al. 1993; Suda et al. 1995). Ligand expression appears to be limited to $CD8^+$ cytotoxic T lymphocytes (CTL) and $CD4^+$, type 1 helper T cells (T_H1) (Rouvier et al. 1993; Ju et al. 1994; Kagi et al. 1994; Ramsdell et al. 1994; Stadler et al. 1994). *In vitro* experiments have demonstrated that CD95L expression by T cells is not constitutive but occurs rapidly following engagement of their antigen receptor. Expression also requires *de novo* synthesis of ligand, is short-lived and can occur in cells which have upregulated CD95 as well (Anel et al. 1994; Luciani and Golstein 1994; Vignaux et al. 1995). In summary, CD95L expression occurs coordinately with CD95 upregulation during an immune response. The significance of these observations is that a T or B cell may have only three or four days to carry out its effector function before it is targeted for destruction by suicide or fratricide/sororicide (Daniel and Krammer 1994; Alderson et al. 1995; Russell and Wang 1993; Brunner et al. 1995; Dhein et al. 1995; Ju et al. 1995). In support of this notion, autoreactive T cells from *lpr* mice constitutively express high levels of functional CD95L as if they are trying to commit suicide but cannot do so as they lack functional CD95 (Chu et al. 1995; Watanabe et al. 1995).

The search for the CD95 ligand was begun by experiments performed by Rouvier and colleagues (1993). This group used a CTL hybridoma (mouse CTL fused to a rat thymoma) designated d10S. When treated with calcium ionophores (ionomycin) and phorbol esters (PMA), this hybridoma kills target cells in an MHC-unrestricted fashion. When mixed with ^{51}Cr-labeled thymocytes, PMA + ionomycin-activated d10S cells were able to kill cells derived from +/+ but not *lpr/lpr* mice. The authors then screened a large panel of murine tumor cells, including YAC-1, EL-4, and P815, before finding that L1210, a DBA/2-derived lymphocytic leukemia cell, was resistant to d10S-mediated killing. Transfection of L1210 with a cDNA encoding mouse CD95 produced a cell, L1210-Fas, which was sensitive to d10S-mediated killing. Nagata's group used the d10S CTL hybridoma to clone CD95L (Suda et al. 1993; Suda and Nagata 1994).

Like membrane-associated tumor necrosis factor, the membrane form of CD95L can be cleaved by a metalloproteinase expressed by many cells types (Kayagaki et al. 1995; Mariani et al. 1995; Tanaka et al. 1996). The soluble molecule can apparently form a trimer which is biologically active, as has been reported for TNF (Smith an Baglioni 1987; Smith et al. 1994). The soluble form of mouse CD95L is unstable *in vitro* and it cannot be detected in culture supernatants of activated mouse lymphocytes (Nagata and Golstein 1995; Takahashi et al. 1994b). In contrast, human lymphocytes have been reported to secrete a stable and functional form of CD95L (Takahashi et al. 1994b; Tanaka et al. 1996).

In addition to performing a role in down-regulation of immune responses, several groups have suggested that CD95/CD95L interactions may play a role in immune-mediated tissue damage (Nagata and Golstein 1995; Ogasawara et al. 1993; Mita et al. 1994; Galle et al. 1995). As described above, CD95 is expressed on many cells of non-hematopoietic origin including liver, heart muscle, ovarian, skin., gut and ductal epithelia (Watanabe-Fukunaga et al. 1992b; Leithauser et al. 1993). Although expressed at high levels on these cells, CD95 appears to be functional only on hepatocytes (Mita et al. 1994; Galle et al. 1995). Mice injected with CD95 antibody die within six hours of massive liver damage without apparent damage to other organs or tissues which express CD95 (Ogasawara et al. 1993; Lacronique et al. 1996). These observations have led to the notion that viral and non-viral hepatitis associated with lymphoid infiltration may be mediated through CD95 in a nonspecific fashion. However, since most immune responses are not associated with hepatotoxicity, it seems likely that a soluble form of CD95L, if it exists *in vivo,* must only be active only very short distances.

2. CD95 LIGAND AND SELF-RECOGNITION BY CYTOTOXIC T LYMPHOCYTES

Our interest in CD95L arose from our long-standing research aimed at understanding how cytotoxic T lymphocytes (CTL) induce apoptosis in their targets (Duke 1992; Duke et al. 1983; Duke and Cohen 1992; Duke et al. 1994). While attempting to repeat an observation made by Lanzavecchia (1986), the assay system which has allowed us to study CD95-dependent killing was discovered. In brief, Lanzavecchia had shown that once activated, cytotoxic T cells can kill any target to which they bind or by which they are bound. We wanted to test the idea that all mature T cells, including CTL, have low but real affinity for "self" MHC, that is MHC which was present in the thymus in which they matured. The experimental design which was developed for examining whether activated CTL can recognize and kill cells bearing self MHC antigens is detailed in Duke (1989). The conclusions of our studies were that triggered, alloreactive CTL clones can effectively kill cells bearing self-MHC antigens. Our results, an example of which are shown in Table 1, provided the first clear demonstration that T cells, or more precisely CTL clones, have detectable affinity for self MHC and thereby supported the notion that T cells are positively selected to mature based on low but real affinity for self MHC in the thymus.

Having discovered that triggered CTL clones were able to induce apoptotic cell death in targets bearing self-MHC we then proceeded to investigate the mechanism of killing mediated by the triggered CTL. One of the first questions which we asked concerned

Table 1. Bystander killing of target cells bearing syngeneic (H-2^d) MHC antigens by "triggered" DAB-6 CTL (DBA/2 anti-C57Bl/6; H-2^d anti-H-2^b)

Target cell	H-2	% specific lysis mediated by DAB-6 without PMA + ionomycin	% specific lysis mediated by DAB-6 pre-incubated with PMA + ionomycin
EL-4	b (allogeneic)	89 ± 3	81 ± 5
CTLL-2	b (allogeneic)	83 ± 2	87 ± 6
P815	d (syngeneic)	1 ± 3	64 ± 3
A20	d (syngeneic)	0 ± 1	58 ± 4
L1210	d (syngeneic)	2 ± 3	12 ± 6
L1210-Fas	d (syngeneic)	5 ± 5	52 ± 3

Methods: DAB-6 CTL were incubated with or without 5 ng/ml PMA and 500 ng/m ionomycin for 2 hr at 37°C to induce CD95L expression (21). 20,000 of the triggered or resting CTL were mixed with 10,000 ^{51}Cr-labeled target cells in individual wells of 96-well V-bottomed plates. The plates were centrifuged at 50 x g for 5 min to establish cell-to-cell contact, placed in the incubator, and % specific lysis was determined after 4 hours.

the involvement of extracellular calcium. We were intrigued to observe that bystander killing by triggered CTL was calcium independent. We wondered if our system was working in a similar fashion to Rouvier's (1993) and obtained L1210 and L1210-Fas (both are H-2^d) from Pierre Golstein in order to answer our question. L1210 is a leukemia cell which lacks expression of CD95 while L1210-Fas has been transfected with murine CD95 (Rouvier et al. 1993). As can be appreciated from the data presented in Table 1, we found that bystander killing by triggered CTL clones was CD95-dependent. We then proceeded to examine whether polyclonal populations of CTL, when triggered, could kill syngeneic T cells expressing CD95.

As discussed above, resting T cells express low levels of the transmembrane CD95 protein. However, within hours of engagement of their antigen receptors, CD95 expression increases significantly (Miyawaki et al. 1992; Owen-Schaub et al. 1992; Alderson et al. 1994; Daniel and Krammer 1994). As shown in Table 2, several days after T cells are first activated, CD95 becomes "functional" and the cells undergo apoptosis in response to CD95 crosslinking with antibody. In contrast, activated T cells from *Lpr* mice, which cannot express functional CD95, are not killed by the antibody.

When CTL generated from C57Bl/6 (B6) mice were stimulated to express CD95L, they killed activated but not resting, syngeneic T cells. In contrast, CTL derived from B6Smn.C3H-*gld* (B6-*gld*) mice, were incapable of killing activated T cells. These observations, together with the observed characteristics of *lpr* and *gld* mice, suggest that expression of functional CD95 on activated lymphocytes targets them for destruction.

Table 2. Triggered CTL derived from mice which express functional CD95L can kill activated T cells which express CD95

Target	B6 CTL			B6-*gld* CTL			Anti-CD95 antibody
	30:1	60:1	120:1	30:1	60:1	120:1	
Resting B6 T	2 ± 2	7 ± 3	12 ± 1	1 ± 0	1 ± 1	4 ± 3	12 ± 5
Activated B6 T	26 ± 3	35 ± 1	47 ± 4	1 ± 1	4 ± 2	8 ± 6	67 ± 8
Activated B6-*lpr* T	5 ± 2	3 ± 2	6 ± 5	2 ± 1	8 ± 7	3 ± 0	9 ± 4

Methods: Alloreactive CTL (B6 or B6-*gld* anti-BALB) were mixed with 5000 ^{51}Cr-labeled target cells (Resting B6 T, freshly isolated B6 lymph node T cells; Activated B6 T, B6 lymph node T cells stimulated with concanavalin A and IL-2 for 4 days; or Activated B6-*lpr* T, B6-*lpr* lymph node T cells stimulated with concanavalin A and IL-2 for 4 days). Prior to mixing, CTL were triggered to express CD95L by pre-incubation in the presence of PMA plus ionomycin (Rouvier et al. 1993). Assays were performed in the presence of 5 mM EGTA so that only CD95L-mediated killing would be detected (Rouvier et al. 1993). Percent specific lysis was calculated after 6 hr.

3. CD95 LIGAND: ROLE IN IMMUNE PRIVILEGE

In addition to activated CD8$^+$ CTL and CD4$^+$ T$_H$1, Nagata and colleagues reported that CD95L mRNA was observed in testis (Suda et al. 1993; Takahashi et al. 1994b; Suda et al. 1995). We have previously reported that rat testis, when placed in the abdominal cavity, is a highly effective immune privileged site into which both allografts and xenografts can be successfully transplanted (Selawry and Whittington 1984; Selawry et al. 1989; Cameron et al. 1990; Selawry and Cameron 1993). We hypothesized that expression of CD95L in testis could account for its immune privileged status.

To test this, a transplant model was employed (Bellgrau et al. 1995). When even weakly immunogenic islet or thyroid tissue from B6 donors was transplanted under the kidney capsule of BALB/c recipients, it was destroyed within two weeks. This is an expected result as BALB/c and B6 mice are genetically incompatible at the major histocompatibility complex (MHC) as well as at multiple minor loci. When similar experiments were performed with testis tissue, we found that B6 testis grafts were maintained and that their gross appearance 21–28 days after transplant into BALB/c recipients was similar to syngeneic grafts. Examination at higher magnification revealed slight evidence of lymphocytic infiltration. These experiments demonstrated that allogeneic testicular tissue, unlike tissue from virtually any other source including pancreatic islet cells, thyroid and skin, was protected from T cell dependent allograft destruction.

In contrast to the results obtained with B6 testis, when B6-*gld* testis tissue was transplanted under the kidney capsule of BALB/c recipients, it was immediately rejected. Essentially no tissue could be found at day 7 after transplantation. Similar experiments using B6.MRL-*lpr* (B6-*lpr*) mice, which express wild-type CD95L but lack functional expression of CD95, revealed that B6-*lpr* testis grafts, like B6, were accepted in BALB/c recipients. B6-*gld* testis did survive when transplanted into CB.17^{scid} (Table 3) and B6-*gld* mice (data not shown) indicating that the failure to engraft in BALB/c mice was not a result of the transplant procedure itself.

Previous studies had indicated that the ability of testis to protect against allograft rejection was afforded by Sertoli cells (Selawry and Cameron 1993). Sertoli cells isolated from B6 mouse testis by enzymatic digestion were transplanted under the kidney capsule of BALB/c recipients. B6 Sertoli cells showed no evidence of rejection when examined 21 days after transplantation. In contrast, Sertoli cell grafts derived from B6-*gld* donors were rejected. A summary of our results is shown in Table 3. These show the effectiveness of

Table 3. CD95L expression protects testis and Sertoli cells from allograft rejection

Donor*	CD95L	Recipient	Challenge[†]	Graft	n	% recovered[§]	% rejected
B6	normal	BALB/c	none	Testis	34	91	0
				Sertoli	12	92	0
B6-lpr	normal	BALB/c	none	Testis	6	100	0
				Sertoli	3	100	0
B6-gld	mutant	BALB/c	none	Testis	34	85	100
				Sertoli	12	67	88
B6-gld	normal	CB.17^{scid}	none	Testis	5	80	0
				Sertoli	5	100	0
B6	normal	BALB/c	B6 spleen	Testis	5	80	0
				Sertoli	2	100	0

* Grafts were placed under the kidney capsule. Mice were sacrificed on days 7-28 after transplant.

[†] Challenged by intraperitoneal injection of 1 x 10^7 spleen cells on day 14 after transplant.

[§] For technical reasons, it was occasionally impossible to recover the graft itself.

Table 4. *Lpr* mice reject CD95L-expressing testis allografts

Donor	CD95L	Recipient	CD95	n	% rejected
BALB/c	normal	B6	normal	5	0
BALB/c	normal	B6-lpr	mutant	8	100

CD95L expression in preventing allograft rejection. In addition, they further show that CD95L expression can prevent graft rejection even in the face of a challenge with donor derived antigen-presenting cells (bottom lines).

To formally demonstrate that Sertoli cells constitutively expressed CD95L transcripts, RT-PCR analysis was performed. As shown previously by Northern analysis (testis (Suda et al. 1993; Takahashi et al. 1994b; Suda et al. 1995), CD95L expression was observed in testis but not in lung. Relevant to the results presented above, CD95L transcripts were detected in freshly isolated Sertoli cells as well as in Sertoli cells cultured at 32°C and 37°C (data not shown) in consideration of the temperatures found in the scrotal vs. abdominal position for transplantation.

In summary, our findings suggest that expression of functional CD95L by Sertoli cells accounts for the immune privileged nature of testis. The link between the failure of B6-*gld* Sertoli cells to protect against allograft immunity (Table 3) and the inability of B6-*gld* T cells to mediate *in vitro* cytotoxicity against activated T cell targets (Table 2) is due to the loss of critical interactions between CD95 and its ligand. Most importantly, and consistent with this model, preliminary studies have shown that BALB/c testis is rejected by B6-*lpr* mice whose graft-specific T cells, when activated, fail to express functional CD95 and are thus able to mediate graft rejection (Table 4).

In addition to protecting against allograft rejection, we have preliminary evidence which suggests that CD95L can prevent xenograft rejection in concordant species. Lewis rats (5 per group) were transplanted under the kidney capsule with one-half of a testis from either B6, B6-*lpr* or B6-*gld* mice. After 28 days, the mice were sacrificed and sections of the transplanted kidneys were examined for histology. B6 grafts were intact and showed only a slight infiltration at the edges. B6-*lpr* grafts were similarly accepted (data not shown). B6-*gld* grafts, in contrast, showed evidence of heavy infiltration and complete loss of donor tissues. Identical to what was demonstrated for allografts, DA rat testis grafts were accepted when transplanted under the kidney capsule of B6 mice but were rejected by B6-*lpr* mice.

Recently, Griffith and colleagues (1995) published results which indicate that CD95-CD95L interactions are an important mechanism for the maintenance of immune privilege in the eye. They showed that inflammatory cells entering the anterior chamber of the eye of normal mice in response to viral infection undergo apoptosis and produced no tissue damage. In contrast, viral infections in *gld* mice resulted in inflammation and invasion of ocular tissue. Furthermore, CD95-expressing but not CD95-negative tumor cells were killed by apoptosis when placed within the anterior chamber of the eyes of normal but not *gld* mice.

4. THE FUTURE: USING CD95L TO CREATE IMMUNE PRIVILEGED TISSUE

Our preliminary results utilizing naturally-occurring, CD95L-mediated immunosuppression suggest that it acts to block only those T cells which are already activated and responding to antigens in close proximity to cells expressing CD95L. CD95L-mediated immunosuppression would be expected to primarily target activated effector cells rather than the activation steps that produce them. The most commonly used immunosuppressive agents, such as cyclosporin A, target T cell lymphokine production but need not block effector function. It is not surprising that these agents might be less effective against previously activated T cells (Hao et al. 1990). By targeting only activated T cells, grafted cell-associated CD95L may provide a highly specific form of immunosuppression for ameliorating T cell-dependent graft rejection and other inflammatory processes (Bellgrau et al. 1995; Vaux 1995).

Towards this goal, Lau and colleagues reported that allogeneic transplantation of islets of Langerhans was facilitated by the co-transplantation of syngeneic muscle cells genetically engineered to express CD95L (Lau et al. 1996; Wickelgren 1996). Composite grafting of allogeneic islets with transfected myoblasts expressing protected islet grafts from immune rejection and maintained normoglycemia for more than 80 days in mice with streptozotocin-induced diabetes. These important results lend strong support to our hypothesis that CD95 ligand could be used to create immune privileged tissue for a variety of transplant purposes.

ACKNOWLEDGMENTS

The authors wish to acknowledge the following organizations and individuals for their generous support: The National Institutes of Health (AI29553; AI34747, AI48808), American Cancer Society (#IRG 5-30), University of Colorado Cancer Center (CA46934, CA49981), Juvenile Diabetes Foundation, Lymphoma Research Foundation, Milheim Foundation, Morrison Charitable Trust, and Theodore and Caroline Shreve. The authors thank Drs. Helena Selawry, Daniel Gold and Pierre Golstein for cells and reagents used in the course of these studies.

REFERENCES

Alderson, M.R., Tough, T.W., Davis-Smith, T., Braddy, S., Falk, B., Schooley, K.A., Goodwin, R.G., Smith, C.A., Ramsdell, F., and Lynch, D.H. 1995. Fas ligand mediates activation-induced cell death in human T lymphocytes. J. Exp. Med. 181, 71–77.

Alderson, M.R., Tough, T.W., Braddy, S., Davis-Smith, T., Roux, E., Schooley, K., Miller, R.E., and Lynch, D.H. 1994. Regulation of apoptosis and T cell activation by Fas-specific mAb. Int. Immunol. 6, 1799–1806.

Anel, A., Buferne, M., Boyer, C., Schmitt-Verhulst, A.M., and Golstein, P. 1994. T cell receptor-induced Fas ligand expression in cytotoxic T lymphocyte clones is blocked by protein tyrosine kinase inhibitors and cyclosporin A. Eur. J. Immunol. 24, 2469–2476.

Bellgrau, D., Gold, D., Selawry, H., Moore, J., Franzusoff, A., and Duke, R.C. 1995. A role for CD95 ligand in preventing graft rejection. Nature. 377, 630–632.

Brunner, T., Mogil, R.J., LaFace, D., Yoo, N.J., Mahboubi, A., Echeverri, F., Martin, S.J., Force, W.R., Lynch, D.H., Ware, C.F., and Green, D.R. 1995. Cell-autonomous Fas (CD95)/Fas-ligand interaction mediates activation- induced apoptosis in T-cell hybridomas. Nature. 373, 441–444.

Cameron, D.F., Whittington, K., Schultz, R.E. and Selawry, H.P. 1990. Successful islet/abdominal testis transplantation does not require Leydig cells. Transplantation 50, 649–653.

Chu, J.L., Ramos, P., Rosendorff, A., Nikolic-Zugic, J., Lacy, E., Matsuzawa, A., and Elkon, K.B. 1995. Massive upregulation of the Fas ligand in lpr and gld mice: implications for Fas regulation and the graft-versus-host disease-like wasting syndrome. J. Exp. Med. 181, 393–398.

Cohen, P.L. and Eisenberg, R.A. 1992. The lpr and gld genes in systemic autoimmunity: life and death in the Fas lane. Immunol. Today. 13, 427–428.

Cohen, P.L. and Eisenberg, R.A. 1991. Lpr and gld: single gene models of systemic autoimmunity and lymphoproliferative disease. Annu. Rev. Immunol. 9, 243–269.

Daniel, P.T. and Krammer, P.H. 1994. Activation induces sensitivity toward APO-1 (CD95)-mediated apoptosis in human B cells. J. Immunol. 152, 5624–5632.

Debatin, K.M., Goldmann, C.K., Bamford, R., Waldmann, T.A., and Krammer, P.H. 1990. Monoclonal-antibody-mediated apoptosis in adult T-cell leukaemia. Lancet. 335, 497–500.

Dhein, J., Walczak, H., Baumler, C., Debatin, K.M., and Krammer, P.H. 1995. Autocrine T-cell suicide mediated by APO-1/(Fas/CD95). Nature. 373, 438–441.

Duke R.C, Chervenak, R., and Cohen, J.J. 1983. Endogenous endonuclease-induced DNA fragmentation: an early event in cell-mediated cytolysis. Proc. Natl. Acad. Sci. USA 80, 6361–6365.

Duke, R.C. 1989. Self-recognition by T cells. I. Bystander killing of target cells bearing syngeneic MHC antigens. J. Exp. Med. 170, 59–71.

Duke, R.C. 1992. Apoptosis in cytotoxic T cells and their targets. Sem. Immunol. 4, 407–412.

Duke, R.C. and Cohen, J.J. 1992. Cell death and apoptosis. In: Current Protocols in Immunology Coligan, J.E., Kruisbeek, A.M., Margulies, D.H., Shevach, E.M. and Strober, W., editors. Greene Publishing Associates, Brooklyn NY, pp. 3.17.1–3.17.16

Duke, R.C., Zulauf, R., Nash, P.B., Zheng, L.M., Young, J.D.E. and Ojcius, D. 1994. Cytolysis mediated by ionophores and pore-forming agents: role of intracellular calcium in apoptosis. FASEB J. 8, 237–246.

Galle, P.R., Hofmann, W.J., Walczak, H., Schaller, H., Otto, G., Stremmel, W., Krammer, P.H., and Runkel, L. 1995. Involvement of the CD95 (APO-1/Fas) receptor and ligand in liver damage. J. Exp. Med. 182, 1223–1230.

Griffith, T.S., Brunner, T., Fletcher, S.M., Green, D.R., and Ferguson, T.A. 1995. Fas ligand-induced apoptosis as a mechanism of immune privilege. Science. 270, 1189–1192.

Hao, L.M., Wang, Y., Gill, R.G., LaRosa, F.G., Talmage, D.W., and Lafferty, K.J. 1990. Role of lymphokines in islet allograft rejection. Transplantation 49, 609–614.

Itoh, N., Yonehara, S., Ishii, A., Yonehara, M., Mizushima, S., Sameshima, M., Hase, A., Seto, Y., and Nagata, S. 1991. The polypeptide encoded by the cDNA for human cell surface antigen Fas can mediate apoptosis. Cell. 66, 233–243.

Ju, S.T., Panka, D.J., Cui, H., Ettinger, R., el-Khatib, M., Sherr, D.H., Stanger, B.Z., and Marshak-Rothstein, A. 1995. Fas(CD95)/FasL interactions required for programmed cell death after T-cell activation. Nature. 373, 444–448.

Ju, S.T., Cui, H., Panka, D.J., Ettinger, R., and Marshak-Rothstein, A. 1994. Participation of target Fas protein in apoptosis pathway induced by CD4+ Th1 and CD8+ cytotoxic T cells. Proc. Natl. Acad. Sci. U. S. A. 91, 4185–4189.

Kagi, D., Vignaux, F., Ledermann, B., Burki, K., Depraetere, V., Nagata, S., Hengartner, H., and Golstein, P. 1994. Fas and perforin pathways as major mechanisms of T cell-mediated cytotoxicity. Science. 265, 528–530.

Kayagaki, N., Kawasaki, A., Ebata, T., Ohmoto, H., Ikeda, S., Inoue, S., Yoshino, K., Okumura, K., and Yagita, H. 1995. Metalloproteinase-mediated release of human Fas ligand. J. Exp. Med. 182, 1777–1783.

Lacronique, V., Mignon, A., Fabre, M., Viollet, B., Rouquet, N., Molina, T., Porteu, A., Henrion, A., Bouscary, D., Varlet, P., Joulin, V., and Kahn, A. 1996. Bcl-2 protects from lethal hepatic apoptosis induced by an anti-Fas antibody in mice. Nat. Med. 2, 80–86.

Lanzavecchia, A. 1986. Is the T-cell receptor involved in T-cell killing? Nature 319, 778–80.

Lau, H.T., Yu, M., Fontana, A., and Stoeckert, C.J. 1996. Prevention of Islet allograft rejection with engineered myoblasts expressing FasL in mice. Science 273, 109–112.

Leithauser, F., Dhein, J., Mechtersheimer, G., Koretz, K., Bruderlein, S., Henne, C., Schmidt, A., Debatin, K.M., Krammer, P.H., and Moller, P. 1993. Constitutive and induced expression of APO-1, a new member of the nerve growth factor/tumor necrosis factor receptor superfamily, in normal and neoplastic cells. Lab. Invest. 69, 415–429.

Luciani, M.F. and Golstein, P. 1994. Fas-based d10S-mediated cytotoxicity requires macromolecular synthesis for effector cell activation but not for target cell death. Philos. Trans. R. Soc. Lond. B. Biol. Sci. 345, 303–309.

Mariani, S.M., Matiba, B., Baumler, C., and Krammer, P.H. 1995. Regulation of cell surface APO-1/Fas (CD95) ligand expression by metalloproteases. Eur. J. Immunol. 25, 2303–2307.

Mita, E., Hayashi, N., Iio, S., Takehara, T., Hijioka, T., Kasahara, A., Fusamoto, H., and Kamada, T. 1994. Role of Fas ligand in apoptosis induced by hepatitis C virus infection. Biochem. Biophys. Res. Commun. 204, 468–474.

Miyawaki, T., Uehara, T., Nibu, R., Tsuji, T., Yachie, A., Yonehara, S., and Taniguchi, N. 1992. Differential expression of apoptosis-related Fas antigen on lymphocyte subpopulations in human peripheral blood. J. Immunol. 149, 3753–3758.

Nagata, S. and Golstein, P. 1995. The Fas death factor. Science. 267, 1449–1456.

Oehm, A., Behrmann, I., Falk, W., Pawlita, M., Maier, G., Klas, C., Li-Weber, M., Richards, S., Dhein, J., Trauth, B.C., Ponsting, H. and Krammer, P.H. 1992. Purification and molecular cloning of the APO-1 cell surface antigen, a member of the tumor necrosis factor/nerve growth factor receptor superfamily. Sequence identity with the Fas antigen. J. Biol. Chem. 267, 10709–10715.

Ogasawara, J., Watanabe-Fukunaga, R., Adachi, M., Matsuzawa, A., Kasugai, T., Kitamura, Y., Itoh, N., Suda, T., and Nagata, S. 1993. Lethal effect of the anti-Fas antibody in mice. Nature. 364, 806–809.

Owen-Schaub, L.B., Yonehara, S., Crump, W.L., and Grimm, E.A. 1992. DNA fragmentation and cell death is selectively triggered in activated human lymphocytes by Fas antigen engagement. Cell Immunol. 140, 197–205.

Ramsdell, F., Seaman, M.S., Miller, R.E., Picha, K.S., Kennedy, M.K., and Lynch, D.H. 1994. Differential ability of Th1 and Th2 T cells to express Fas ligand and to undergo activation-induced cell death. Int. Immunol. 6, 1545–1553.

Roths, J.B., Murphy, E.D., and Eicher, E.M. 1984. A new mutation, gld, that produces lymphoproliferation and autoimmunity in C3H/HeJ mice. J. Exp. Med. 159, 1–20.

Rouvier, E., Luciani, M.F., and Golstein, P. 1993. Fas involvement in Ca(2+)-independent T cell-mediated cytotoxicity. J. Exp. Med. 177, 195–200.

Russell, J.H. and Wang, R. 1993. Autoimmune gld mutation uncouples suicide and cytokine/proliferation pathways in activated, mature T cells. Eur. J. Immunol. 23, 2379–2382.

Selawry, H.P. and Cameron, D.F. 1993. Sertoli cell-enriched fractions in successful islet cell transplantation. Cell Transplant. 2, 123–129.

Selawry, H.P. and Whittington, K. 1984. Extended allograft survival of islets grafted into intra-abdominally placed testis. Diabetes 33, 405–406.

Selawry, H.P., Whittington, K.B. and Bellgrau, D. 1989. Abdominal intratesticular islet-xenograft survival in rats. Diabetes 38 Suppl 1, 220–223.

Smith, C.A., Farrah, T., and Goodwin, R.G. 1994. The TNF receptor superfamily of cellular and viral proteins: activation, costimulation, and death. Cell. 76, 959–962.

Smith, R.A. and Baglioni, C 1987. The active form of tumor necrosis factor is a trimer. J. Biol. Chem. 262, 6951–6954.

Sobel, E.S., Kakkanaiah, V.N., Kakkanaiah, M., Cohen, P.L., and Eisenberg, R.A. 1995. Co-infusion of normal bone marrow partially corrects the gld T-cell defect. Evidence for an intrinsic and extrinsic role for Fas ligand. J. Immunol. 154, 459–464.

Sobel, E.S., Kakkanaiah, V.N., Cohen, P.L., and Eisenberg, R.A. 1993. Correction of gld autoimmunity by co-infusion of normal bone marrow suggests that gld is a mutation of the Fas ligand gene. Int. Immunol. 5, 1275–1278.

Stalder, T., Hahn, S., and Erb, P. 1994. Fas antigen is the major target molecule for CD4+ T cell-mediated cytotoxicity. J. Immunol. 152, 1127–1133.

Strasser, A. 1995. Apoptosis. Death of a T cell. Nature. 373, 385–386.

Suda, T., Okazaki, T., Naito, Y., Yokota, T., Arai, N., Ozaki, S., Nakao, K., and Nagata, S. 1995. Expression of the Fas ligand in cells of T cell lineage. J. Immunol. 154, 3806–3813.

Suda, T. and Nagata, S. 1994. Purification and characterization of the Fas-ligand that induces apoptosis. J. Exp. Med. 179, 873–879.

Suda, T., Takahashi, T., Golstein, P., and Nagata, S. 1993. Molecular cloning and expression of the Fas ligand, a novel member of the tumor necrosis factor family. Cell. 75, 1169–1178.

Takahashi, T., Tanaka, M., Brannan, C.I., Jenkins, N.A., Copeland, N.G., Suda, T., and Nagata, S. 1994a. Generalized lymphoproliferative disease in mice, caused by a point mutation in the Fas ligand. Cell. 76, 969–976.

Takahashi, T., Tanaka, M., Inazawa, J., Abe, T., Suda, T., and Nagata, S. 1994b. Human Fas ligand: gene structure, chromosomal location and species specificity. Int. Immunol. 6, 1567–1574.

Tanaka, M., Suda, T., Haze, K., Nakamura, N., Sato, K., Kimura, F., Motoyoshi, K., Mizuki, M., Tagawa, S., Ohga, S., Hatake, K., Drummond, A.H., and Nagata, S. 1996. Fas ligand in human serum. Nat. Med. 2, 317–322.

Trauth, B.C., Klas, C., Peters, A.M., Matzku, S., Moller, P., Falk, W., Debatin, K.M., and Krammer, P.H. 1989. Monoclonal antibody-mediated tumor regression by induction of apoptosis. Science. 245, 301–305.

Vaux, D.L. 1995. Immunology. Ways around rejection. Nature. 377, 576–577.

Vignaux, F., Vivier, E., Malissen, B., Depraetere, V., Nagata, S., and Golstein, P. 1995. TCR/CD3 coupling to Fas-based cytotoxicity. J. Exp. Med. 181, 781–786.

Vignaux, F. and Golstein, P. 1994. Fas-based lymphocyte-mediated cytotoxicity against syngeneic activated lymphocytes: a regulatory pathway?. Eur. J. Immunol. 24, 923–927.

Watanabe, D., Suda, T., Hashimoto, H., and Nagata, S. 1995. Constitutive activation of the Fas ligand gene in mouse lymphoproliferative disorders. EMBO J. 14, 12–18.

Watanabe-Fukunaga, R., Brannan, C.I., Itoh, N., Yonehara, S., Copeland, N.G., Jenkins, N.A., and Nagata, S. 1992b. The cDNA structure, expression, and chromosomal assignment of the mouse Fas antigen. J. Immunol. 148, 1274–1279.

Watanabe-Fukunaga, R., Brannan, C.I., Copeland, N.G., Jenkins, N.A., and Nagata, S. 1992a. Lymphoproliferation disorder in mice explained by defects in Fas antigen that mediates apoptosis. Nature. 356, 314–317.

Wickelgren, I. 1996. Muscling transplants into mice. Science 273, 35.

Yonehara, S., Ishii, A., and Yonehara, M. 1989. A cell-killing monoclonal antibody (anti-Fas) to a cell surface antigen co-downregulated with the receptor of tumor necrosis factor. J. Exp. Med. 169, 1747–1756.

INDEX

ACAID, 154
AE7 effector cells, 81
AICD, 107, 149–152
Amphibian metamorphosis, 1–9, 13–23
Angiogenesis, 35–42
Anti-lg/slg, 79–86, 88
Apo-1, 159–165
Apoptosis, 63, 79–86, 88, 113–122
Aspase, 45, 46
Autoantibody, 84–88
Autoimmunity, 79, 88
Autoreactive, 84–88
A23187, 137–138

Baculovirus, 35, 52, 60–61
BAF/BO3, 120–121
Basement membrane, 27, 28, 30–33
Bax, 39, 72
B cells, 79–88
Bcl-2 family, 114
    Bad, 119
    Bak, 119
    Bax, 119–120
    Bcl-2, 7, 39, 114, 119–121
    Bcl-x, 72, 119
    Mcl, 119
Bcl-X$_L$, 82, 86
Blk, 93
Blood vessel formation, 35
Boc-asp-FMK, 53, 56–58, 59–60

Castration, 31, 32
CD3, 105
CD3$^{+/+}$, 106
CD3$^{-/-}$, 106
CD28, 115
CD40, 99, 115
CD8-E chimera, 109
CD95, 159–165
CD40/CD40L, 79–86, 88
CDK, 72
Ced3, 46
Cell adhesion, 37

Cell cycle, 37, 113–122
Cell shrinkage, 63
Cell volume, 64, 67
Ceramide, 82
Chromiun release, 81, 82
CH31 lymphoma, 93, 97, 100
Clusterin, 31
C-Myc, 100–101, 142–143
Collagen, 40
Collagenase, 28
CpG motif, 96
CPP32, 46–48, 140
Cyclin dependent kinase 2 (cdk2), 96
Cyclin A, 96
Cytokine, 113–122
Cytokine-promoted apoptosis, 113, 115
    EGF, 116
    IGF, 116
    IL-2, 116, 120
    PDGF, 116
    TNF, 116
Cytotoxic, 71–77
Cytotoxic T lymphocyte (CTL), 58–61, 160–162

Denatured collagen, 40
Dexamethasone, 4,6, 137–140
Differential display, 82, 86
Differential screen, 17
DNA damage, 71–77
DNA degradation, 67
DNA fragmentation, 45–49
Dominant negative mutant, 76

Egr-1, 142
Endothelial cells, 37–41
Extracellular matrix (ECM), 13–23, 27–29, 31, 32,
    36–41

Fas (CD95), 47, 79–88, 97–99, 147–156,159–165
Fas ligand (CD95L), 29, 31–33, 79–88, 147–156,
    159–165
FLICE, 47